AQA GCSE (9–1)
Biology

Grade 6–7 Booster Workbook

Mike Smith
Shaista Shirazi

William Collins' dream of knowledge for all began with the publication of his first book in 1819. A self-educated mill worker, he not only enriched millions of lives, but also founded a flourishing publishing house. Today, staying true to this spirit, Collins books are packed with inspiration, innovation and practical expertise. They place you at the centre of a world of possibility and give you exactly what you need to explore it.

Collins. Freedom to teach.

Published by Collins
An imprint of HarperCollins*Publishers*
The News Building
1 London Bridge Street
London
SE1 9GF

HarperCollins*Publishers* Macken House
39/40 Mayor Street Upper
Dublin 1
DO1 C9W8
Ireland

Browse the complete Collins catalogue at
www.collins.co.uk

© HarperCollins*Publishers* Limited 2019

10 9 8 7 6 5 4 3

ISBN 978-0-00-832254-0

British Library Cataloguing-in-Publication Data
A catalogue record for this publication is available from the British Library.

Authors: Mike Smith, Shaista Shirazi
Development editor: Elizabeth Barker
Commissioning editors: Rachael Harrison and Jennifer Hall
In-house editor: Joanna Ramsay
Copyeditor: Elizabeth Barker
Proof reader: Helen Bleck
Answer checker: Jouve India Private Limited and Amanda Harman
Cover designers: The Big Mountain Design & Creative Direction
Cover photos: Shutterstock/JuliusKielaitis, Respiro/Shutterstock
Typesetter: Jouve India Private Limited
Illustrators: Jouve India Private Limited
Production controller: Katharine Willard
Printed and bound by: Ashford Colour Press Ltd

The publishers gratefully acknowledge the permission granted to reproduce the copyright material in this book. Every effort has been made to trace copyright holders and to obtain their permission for the use of copyright material. The publishers will gladly receive any information enabling them to rectify any error or omission at the first opportunity.

Contents

Introduction

This workbook will help you build your confidence in answering Biology questions for AQA GCSE Biology and GCSE Combined Science.

It gives you practice in using key scientific words, writing longer answers and answering synoptic questions, as well as applying knowledge and analysing information.

You will find all the different question types in the workbook so you can get plenty of practice in providing short and long answers.

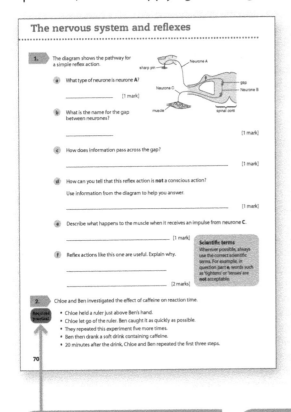

Higher Tier only content is clearly marked throughout.

The questions also cover required practicals, maths skills and synoptic questions – look out for the tags which will help you identify these questions.

Learn how to answer test questions with annotated worked examples. This will help you develop the skills you need to answer questions.

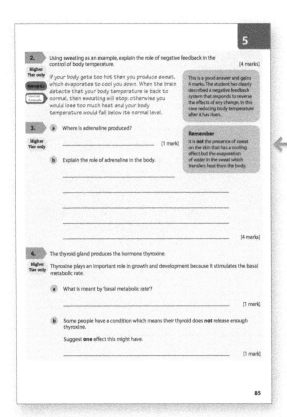

There are lots of hints and tips to help you out. Look out for tips on how to decode command words, key tips for required practicals and maths skills, and common misconceptions.

As you get further through the workbook you should be able to complete more of the questions without the support of the hints and tips.

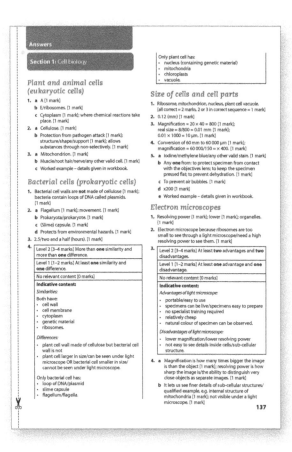

There are answers to all the questions at the back of the book. You can check your answers yourself or your teacher might tear them out and give them to you later to help you mark your work.

Plant and animal cells (eukaryotic cells)

1. The diagram shows an animal cell.

a Which letter shows the structure responsible for controlling substances entering and leaving the cell?

_____ [1 mark]

b Which structure(s) can only be seen with an electron microscope?

_____ [1 mark]

c Identify part **C**. What is the function of part **C**?

Part **C**: _____ [1 mark]

Function: _____ [1 mark]

2. **a** What are plant cell walls made of?

_____ [1 mark]

b State **three** functions of a plant cell wall.

1 _____ [1 mark]

2 _____ [1 mark]

3 _____ [1 mark]

3. **a** Name this organelle. _____ [1 mark]

length

b Write down the name of **one** type of cell where many of these organelles are found.

_____ [1 mark]

Maths

c The organelle has been magnified 12 000 times.

Calculate the actual length of this organelle in μm. [3 marks]

Worked Example

Length on diagram = 15 mm

$$\text{real size} = \frac{\text{size of image}}{\text{magnification}} = \frac{15}{12\ 000}$$

$$= 0.00125 \text{ mm}$$

$$= 1.25 \text{ μm}$$

The student gets all 3 marks for this answer. They have measured the length of the image and used the correct formula to work out the real size of the organelle. They have converted mm to μm correctly by multiplying by 1000 – because there are 1000 μm in 1mm.

Bacterial cells (prokaryotic cells)

1. Which of the following statements are true? Tick **two** boxes.

☐ Bacteria are easily seen under a light microscope with low power.

☐ Bacterial cell walls are **not** made of cellulose.

☐ Bacterial cells contain DNA in a nucleus.

☐ Bacteria contain loops of DNA called plasmids. [2 marks]

2. The diagram shows a bacterial cell.

a Name structure **A** and state its function.

Structure **A**: _____ [1 mark]

Function: _____ [1 mark]

b Which group of organisms do bacteria belong to?

_____ [1 mark]

c What is the outside coating on some bacterial cell walls?

_____ [1 mark]

d Why is this coating important?

_____ [1 mark]

3. In a culture of bacteria, the initial concentration is 2000 cells/cm³.

Maths Each cell divides once every 30 minutes.

Calculate how long it will take for the concentration to become greater than 60 000 cells/cm³.

_____

_____

Answer = _____ hours   [1 mark]

4. Compare a bacterial cell with a plant cell.

_____

_____

_____

_____

_____   [4 marks]

Size of cells and cell parts

1. Write the following cell parts in order of size starting with the smallest.

nucleus   mitochondrion   plant cell vacuole   ribosome

_____ smallest

_____

_____

_____ largest   [2 marks]

2. An egg cell is one of the largest human cells, with an average diameter of 120 μm. Write this in mm.

Maths

Answer _____ mm   [1 mark]

Maths
Converting units can be confusing. Remember there are 1000 micrometres (μm) in every millimetre (mm). So, to convert μm to mm you need to divide by 1000 because you are converting a smaller unit to a bigger one.

3.

Maths

Emily used an eyepiece lens with a magnification of ×20 and an objective lens with a magnification of ×40 to look at a specimen.

The image size of the specimen is 8 mm.

Calculate the real size of the specimen. Write your answer in micrometres.

_____

_____

_____

Answer _____ μm   [3 marks]

4.

Maths

Alex measured the distance between the two ends of a plant cell on a picture. She found the length was 60 mm. The actual length of the cell was 150 μm.

What is the magnification of the cell picture?

_____

_____

Magnification = _____   [2 marks]

5.

Required practical

Ali looked at some onion cells under a light microscope.

a He used a dye to stain the cells. Suggest the name of the dye.

_____   [1 mark]

b He used a coverslip over the specimen. Explain why.

_____

_____   [1 mark]

c Ali tried to lower the coverslip carefully, putting one side down first before the other. Suggest why.

_____

_____   [1 mark]

d Ali used a ×10 eyepiece lens with a ×20 objective lens to view the onion cells. What was the total magnification?

_____

Magnification = _____ [1 mark]

Maths

e Ali calculated the length of one onion cell to be 150 μm. Express this in standard form. [2 marks]

Higher Tier only

Worked Example

$1 \text{ μm} = 0.000001 \text{ m} = 1 \times 10^{-6} \text{ m}$

$150 \text{ μm} = 150 \times (1 \times 10^{-6}) = 1.5 \times 10^{-4} \text{ m}$

So, the onion cell is 1.5×10^{-4} m long.

This answer gains full marks.

Maths

When converting measurements in mm or μm to sizes in metres we often use standard form to avoid writing many zeros after the decimal point.

Electron microscopes

1. Use the correct words from the box to complete the sentences.

| colour range | higher | lower | molecules |
| organelles | organs | resolving power | similar |

An electron microscope has a greater _____ than a light microscope.

A light microscope has a _____ magnification compared with an electron microscope.

An electron microscope can show fine details of _____ in a cell. [3 marks]

2. Name the type of microscope needed to see ribosomes. Explain your answer.

Type of microscope: _____

Explanation: _____

_____ [1 mark]

3. Describe the advantages and disadvantages of using a light microscope compared with using an electron microscope.

_____

_____

_____

_____

_____ [4 marks]

4. **a** Explain the difference between magnification and resolving power.

_____

_____ [2 marks]

Remember

Think of a digital camera image. You can magnify the image to any size you want; but at some point you will lose clarity (resolving power).

b How has electron microscopy increased our understanding of sub-cellular structures?

_____

_____ [2 marks]

Growing microorganisms

1. Bacteria multiply by cell division. What is the name of this type of cell division? Tick **one** box.

☐ Binary fission

☐ Differentiation

☐ Mitosis

☐ Specialisation [1 mark]

2. Sophia investigated the effect of three different antibiotics on bacterial growth. The figure shows the results.

a Which antibiotic is **most** effective at preventing the growth of bacteria? Explain.

Antibiotic: _____

Explanation: _____

_____ [2 marks]

b Sophia wants to compare antiseptics using the same method. Write down **two** things she must do to make the experiment repeatable.

1 _____ [1 mark]

2 _____ [1 mark]

c What is the purpose of the agar? Tick **one** box.

☐ To prevent evaporation.

☐ To provide nutrients for the bacteria.

☐ To provide oxygen for the bacteria.

☐ To support the discs. [1 mark]

3. The radius of the zone of inhibition for antibiotic A was 20 mm. Calculate the area of the zone of inhibition giving your answer in cm^2.

The zone of inhibition includes the area of the paper disc.

_____

_____

_____

Answer = _____ cm^2 [3 marks]

Cell specialisation and differentiation

1. **a** Identify these specialised cells.

A

B

A: _____ [1 mark]

B: _____ [1 mark] C

C: _____ [1 mark]

D

D: _____ [1 mark]

b Explain how cell **D** is specialised for its function.

_____

_____ [2 marks]

c What substance does cell **A** carry around the body?

_____ [1 mark]

2. This epithelial cell is found in the lining of the small intestine.

A _____

a What is the function of this cell?

B _____

_____ [1 mark]

b Write down the names of the structures labelled **A** and **B** on the diagram above.

[2 marks]

c What are the functions of the parts you have labelled in part **b**?

A: _____ [1 mark]

B: _____ [1 mark]

3. What is differentiation? Explain the difference between differentiation in plant and animal cells.

_____

_____

_____

_____

_____

Literacy
Make sure you understand the difference between cell differentiation and cell specialisation. Both terms are closely linked and are often confused by students.

[4 marks]

Cell division by mitosis

1. Complete the sentences using words from the box. The words can be used once, more than once or not at all.

daughter	differentiation	meiosis	mitosis	parent	sister

A _____ cell divides to produce two new _____ cells.

Each _____ cell is genetically identical to the other and to the

_____ cell. This type of cell division is called _____ . [4 marks]

2. Which of the following statements are true? Tick **two** boxes.

☐ Chromosomes are made of a chemical called DNA.

☐ Different types of organism have different numbers of chromosomes in their cells.

☐ Human body cells have 46 pairs of chromosomes.

☐ The genetic information of an organism is contained in the ribosomes.   [2 marks]

3. Before a cell can divide, it needs to grow and increase the number of sub-cellular structures taking part in the process. Name **two** sub-cellular structures taking part in this process.

_____ and _____   [2 marks]

4. Mia made a slide of a garlic root tip. She counted the number of cells in each stage of the cell cycle. The table shows her results.

	Stages in the cycle			
	Stage 1	**Stage 2**	**Stage 3**	**Total**
Number of cells	190	30	20	240

a Which stage of the cell cycle is the shortest?

_____

[1 mark]

Maths **b** What percentage of cells are in this stage? Give your answer to 2 significant figures.

_____

Answer = _____ % [2 marks]

Maths **c** The cell cycle in a garlic root tip cell lasts 12 hours. Calculate the length of time Stage 2 lasts in a typical cell. Give your answer in minutes.

_____

_____

Maths

The fraction of cells in a particular stage of the cell cycle is proportional to the time taken for that stage.

Time in Stage 2 = _____ minutes [2 marks]

5. Describe the three stages of the cell cycle.

_____

_____

_____

_____

_____

_____

_____ [6 marks]

Stem cells

1. What is a stem cell? Explain why stem cells are important for an organism's growth.

_____

_____ [2 marks]

2. Scientists want to make human stem cells from body cells rather than getting stem cells from embryos. Why? Tick **one** box.

☐ Human embryos are single-celled.

☐ Some people object to destroying human embryos.

☐ Stem cells cannot be found in human embryos.

☐ The cells in human embryos are all differentiated. [1 mark]

3. **a** Where are plant stem cells found?

_____ [1 mark]

b Give **one** way that stem cells from human embryos differ from stem cells from plants.

_____

_____ [1 mark]

4. **a** People who disagree with using stem cells from an embryo may **not** disagree with using stem cells from an umbilical cord. Suggest a reason for this.

_____

_____

[1 mark]

Command word

The command word 'suggest' means applying your knowledge to a new situation and, in this case, to give a possible explanation.

b Stem cells from umbilical cords can be stored for future use. Suggest **one advantage** and **one disadvantage** of using stem cells from a person's own umbilical cord to treat a health condition they develop in later life.

Advantage: _____

_____ [1 mark]

Disadvantage: _____

_____ [1 mark]

5. What are the risks and benefits of using stem cells for medical treatments?

_____

_____

_____

_____

_____

_____ [4 marks]

Diffusion in and out of cells

1. Explain what is meant by diffusion. [2 marks]

When the gradient is higher, substances move by diffusion.

> This answer gains 0 marks. The word 'concentration' is key here and just writing 'gradient' does **not** gain credit. There must be an adequate explanation that diffusion only operates down a concentration gradient and that there is net movement from an area of higher concentration to an area of lower concentration. To get full marks, you should mention molecules, atoms, ions, or particles rather than just substances.

2. The diagram shows an animal cell surrounded by oxygen molecules.

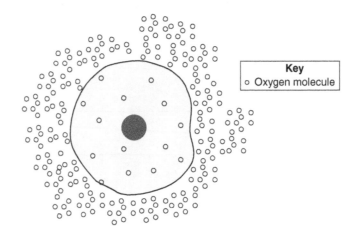

Key
∘ Oxygen molecule

a In which direction will oxygen diffuse? Draw an arrow on the diagram to show your answer.

Explain why the oxygen will move in this direction.

_____ [1 mark]

b The rate of diffusion occurs faster if the temperature is increased.

Give **one** other condition that helps to increase the rate of diffusion.

_____ [1 mark]

3. The diagram shows three cells, **A**, **B** and **C**, surrounded by blood with a 30% oxygen concentration. Each cell contains a different concentration of oxygen.

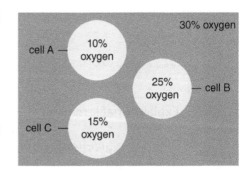

a Into which cell will oxygen move fastest?

_____ [1 mark]

b Explain why.

_____

_____ [1 mark]

4. Explain why diffusion is important to animals and plants. In your answer, use examples of the diffusion of named substances.

_____

_____

_____

_____

_____

_____ [6 marks]

Exchange surfaces in animals

1. The cells from the lining of the lungs are specialised to allow gases to pass through quickly. Give **one** feature of these cells that allows oxygen to pass through quickly.

_____ [1 mark]

2. The figure below is a single alveolus surrounded by a capillary.

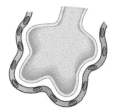

> **Remember**
> Be precise about where you place arrows, particularly where they start and end.

On the diagram, draw and label **one** arrow to show the path taken by oxygen and **one** arrow to show the path taken by carbon dioxide. [2 marks]

3. Jon investigated the effect of size on the uptake of substances by diffusion. He cut differently sized cubes of agar and placed them in beakers of dye. He timed how long it took for the dye to diffuse to the centre of each cube.

The table shows the relationship between surface area and volume of the cubes.

Cube number	Cube size (cm)	Surface area (cm²)	Volume (cm³)	Ratio of surface area : volume
1	2 × 2 × 2	24	8	3 : 1
2	3 × 3 × 3	A	B	2 : 1
3	5 × 5 × 5	C	125	D

a Calculate the values of **A**, **B**, **C** and **D** in the table.

A = _____ cm²   [1 mark]

B = _____ cm³   [1 mark]

C = _____ cm²   [1 mark]

D = _____   [1 mark]

b Which cube took the longest time to change colour?

_____   [1 mark]

c Explain your answer to part **b**.

_____   [1 mark]

4. Explain **three** adaptations of the gills of a fish for efficient gas exchange.

_____

_____

_____

_____

_____

_____

_____

_____   [6 marks]

Osmosis

1. **a** Name the process in which water moves into and out of cells.

_____ [1 mark]

b Water moves through a *partially* permeable membrane. Explain what a *partially* permeable membrane is.

_____

_____ [1 mark]

2. Plant tissue was put into two different strengths of sugar solution, **A** and **B.** Sugar solution **A** is weak while sugar solution **B** is a strong sugar solution.

a Describe **two** ways in which cells in the two tissues will look different from each other after an hour.

1 _____

_____ [1 mark]

2 _____

_____ [1 mark]

b Suggest and describe what happens to red blood cells if they are put in pure water.

_____

_____ [2 marks]

3. The three cells below are surrounded by pure water. The diagram shows the percentage concentration of sugar solution in each cell.

cell A — 2% sugar solution

5% sugar solution — cell B

cell C — 10% sugar solution

Into which cell will water move fastest?
Explain why. [3 marks]

Worked Example

Cell A. Because it has the lowest concentration of water.

4. Some students investigated the effect of different concentrations of sugar on the movement of water.

Required practical

The diagram shows how they set up the investigation.

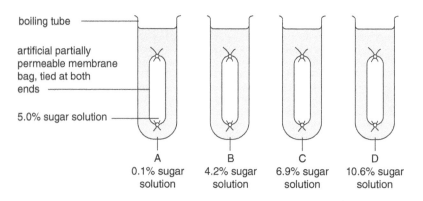

boiling tube

artificial partially permeable membrane bag, tied at both ends

5.0% sugar solution

A
0.1% sugar solution

B
4.2% sugar solution

C
6.9% sugar solution

D
10.6% sugar solution

The students weighed each bag before placing it in one of the boiling tubes **A**, **B**, **C** and **D**. They removed the bags after 30 minutes and weighed them again.

a The bag in boiling tube **A** was heavier after 30 minutes.

Explain why.

_____

_____

_____

_____ [3 marks]

b In which boiling tube, **A**, **B**, **C** or **D**, would you expect the bag to show the smallest change in mass?

_____ [1 mark]

c Explain why you think the bag chosen in part b would show the smallest change.

_____

_____

_____

_____

[2 marks]

d The students decided to calculate the **percentage** change in mass for the sugar solution in the bags for each boiling tube. Why is the percentage change in mass more useful than just the change in mass in grams?

_____

_____

_____ [2 marks]

Misconception

A common misconception is that water moves from a high sugar concentration to a lower sugar concentration. For osmosis questions, you need to think about the **water** concentration in a solution. For example, a 0.1% sugar solution has a higher water concentration than a 5.0% sugar solution. Therefore, water would move from the 0.1% solution in boiling tube **A** into the solution in the bag.

Literacy

Many students struggle to explain terms clearly. For example, when talking about concentration, 'amount of sugar' results in no credit because it does not show an understanding of the concentrations on **both** sides of the permeable membrane.

Active transport

1. **a** Name **one** mineral ion that plants absorb through their roots.

_____ [1 mark]

b Name **two** ways that active transport is different from diffusion.

1 _____ [1 mark]

2 _____ [1 mark]

2. The table shows the concentration of four mineral ions outside and inside plant root hair cells.

Mineral ion	Concentration outside cells in mmol per dm³	Concentration inside cells in mmol per dm³
Chloride	127	5
Magnesium	5	43
Potassium	9	148
Sodium	135	12

Use information from the table to complete the following sentences.

a Chloride ions move into cells by the process of _____ . [1 mark]

b Magnesium ions move into cells by the process of _____ . [1 mark]

c Why is it important that minerals move into root hair cells?

_____

_____ [1 mark]

d Give **one** adaptation of the root hair cell to its function. Explain why the adaptation is important.

_____

_____ [2 marks]

3. **a** Where does active transport occur in the digestive system?

_____ [1 mark]

b Name **one** food molecule that is absorbed by active transport.

_____ [1 mark]

c Explain why it is necessary to absorb some food molecules by active transport.

_____

_____ [2 marks]

Digestive system

1. The diagram shows the digestive system.

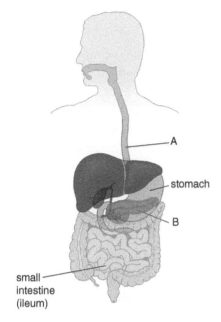

    A

    stomach

    B

small intestine (ileum)

   a What is the name of organ **A**?

    _____ [1 mark]

   b What is the name of organ **B**?

    _____ [1 mark]

2. **a** Name **one** chemical that the stomach produces.

    _____ [1 mark]

   b What is the role of this chemical?

    _____ [1 mark]

3. Describe what is meant by an 'organ'.

    _____

    _____ [2 marks]

4. How does food move along the digestive system?

_____ [1 mark]

5. Bile is released and mixed with food that has left the stomach.

Explain where the bile comes from and why it is needed. [6 marks]

Worked Example

Bile is produced in the liver. It helps to digest fats by breaking large droplets into smaller droplets. It is also alkaline so it neutralises any hydrochloric acid present.

> This answer would gain 3 marks. Although everything the student has said is correct, they need to give more detail to gain full marks. They should have realised this from the number of marks and the number of answer lines that would have been available. Although bile is produced in the liver, they should have mentioned that it is then stored in the gall bladder before being released. They should also explain why it's necessary to break large fat droplets into smaller ones. The reason is that this then provides a larger surface area for the fat-digesting enzyme lipase to work on. If possible, always try to use the correct scientific term: the term for breaking down fat droplets is emulsification.

6. Coeliac disease is a disease affecting the small intestine.

It causes villi to become flattened or even disappear altogether.

People with coeliac disease are often underweight. Suggest why.

Command words

The word 'suggest' is used because this is not a disease that you need to know about. Instead, use your knowledge about the function of villi in a healthy digestive system to answer the question.

_____

_____

_____

_____ [4 marks]

Digestive enzymes

1. Using words from the box, complete the sentences.

carbohydrate	catalyst	fat
glycerol	hormone	liver
salivary glands	stomach	sugar

An enzyme is a _____. [1 mark]

The enzyme protease is made in the _____. [1 mark]

The enzyme lipase breaks down _____ molecules. [1 mark]

2. Describe how carbohydrates are digested to simple sugars. In your answer, include the enzymes involved and the parts of the digestive system involved.

_____

_____

_____

_____

_____ [4 marks]

3. Describe how to test if a lemonade drink contains sugar.

Required practical

_____

_____

_____

_____

_____ [3 marks]

> **Remember**
>
> In your exam you could be asked how to carry out tests to detect the presence of sugar, starch, protein or lipids, and the colour changes to expect for a positive test. Think of ways to remember the colour changes, e.g. the Biuret test for protein is purple if positive.

Factors affecting enzymes

1. Sam and Ada investigated the effect of pH on the digestion of starch by amylase.

Required practical

They tested five different pH buffer solutions.

In the experiment they:

- put amylase and starch solutions in separate test tubes in a beaker of water at 25 °C
- put single drops of iodine in wells on a spotting tile
- added to another test tube 2 cm³ of amylase, 1 cm³ of pH buffer and finally 2 cm³ of starch; then they started timing

- removed a drop of the mixture, after 10 seconds, to one of the wells in the spotting tile
- repeated this step every 10 seconds until the iodine solution remained orange
- repeated this procedure with all the remaining four pH buffers.

a Explain why they timed until the iodine solution remained orange.

_____

_____ [1 mark]

b Explain why they kept the temperature at 25 °C.

_____

_____ [1 mark]

The table shows their results.

pH of solution	Time for colour change to occur in seconds			
	Test 1	Test 2	Test 3	Mean
5	140	150	150	147
6	60	70	70	67
7	30	30	30	30
8	80	80	70	
9	140	140	80	

c Calculate the missing mean values, taking account of any anomalous results.

Write your answers in the table. [2 marks]

d Look at the table.

Describe the effect of pH on the digestion of starch by amylase.

_____

_____

_____

_____ [3 marks]

e Explain why pH affects an enzyme-controlled reaction.

_____

_____

_____

_____ [3 marks]

f Explain **one** way Sam and Ada could improve the method to find more accurate values for this investigation.

_____

_____ [2 marks]

The heart and blood vessels

1. This is a diagram of the heart.

a Name the structures labelled **A** and **B**.

A: _____ [1 mark]

B: _____ [1 mark]

b Where does deoxygenated blood travel through the heart?

Show your answer by drawing arrows on the diagram. [1 mark]

2. The pacemaker cells are found in the heart.

a Where in the heart are pacemaker cells located?

_____ [1 mark]

b Describe the function of the pacemaker cells.

_____ [1 mark]

3. The heart is part of a double circulatory system.

What is meant by a 'double circulatory system'?

_____

_____

_____ [2 marks]

4. Describe and explain **one** way that capillaries are adapted to their function.

_____

_____

_____ [3 marks]

5. 1600 ml of blood moves through an artery in 4 minutes. Calculate the rate of blood flow through the artery, in l/min.

Maths

_____

_____

Rate of blood flow = _____ l/min   [2 marks]

> **Maths**
>
> You will be expected to know how to carry out rate calculations for the flow of blood. The formula will not necessarily be provided, so make sure you know the formula: rate of flow of blood = volume of blood / time taken.

6. Describe and explain the differences between arteries and veins.   [6 marks]

Worked Example

Arteries carry blood at high pressure away from the heart, so they need to have thick walls containing muscle and elastic tissue to withstand and maintain the pressure. Veins carry blood at low pressure back to the heart, so they have thinner walls than arteries, as they don't need to be so thick. They also have a wide lumen to make it easier for blood to flow, and valves to stop blood flowing backwards.

> This is a good answer, describing the differences between arteries as well as explaining them, and would be awarded 5 out of 6 marks. From the structure of the answer, an examiner would assume that 'They' in the last sentence refers to veins, but it would have been better if the student had written 'Veins' instead to make it absolutely clear. If you get an answer asking you to compare two or more things, try to avoid using words like 'they' or 'it' and instead be clear what you are writing about. Unless it is absolutely clear to the examiner what your answer is referring to you may lose marks.

Blood

1. Name **two** substances that are transported in blood plasma.

1 _____ [1 mark]

2 _____ [1 mark]

2. **a** A person has 2640 cm³ of plasma in their body out of a total blood volume of 4800 cm³. What percentage of their blood is plasma?

Maths

Maths
When answering a Maths question, always show your working out, even if you are not asked for it. Even if you get the final answer wrong, you may still gain some marks for your working.

_____

Percentage = _____% [1 mark]

b What usually makes up most of the rest of blood?

_____ [1 mark]

3. Some medical conditions cause low white blood cell counts. This means that the number of white blood cells is lower than usual.

Explain why a doctor would be concerned about a low white blood cell count.

_____

_____

_____ [2 marks]

4. Describe and explain **three** ways in which red blood cells are adapted to transport oxygen.

1 _____

_____ [2 marks]

2 _____

_____ [2 marks]

3 _____

_____ [2 marks]

Heart and lungs

1. This is a diagram of the lungs.

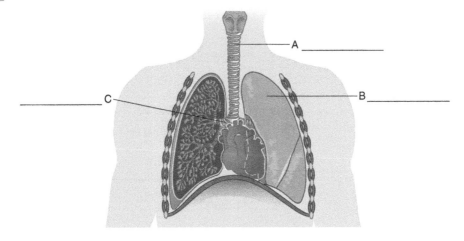

a) Label parts **A**, **B** and **C** on the diagram. [3 marks]

b) Describe and explain how **one** feature of part **A** helps it carry out its function.

Feature: _____ [1 mark]

Explanation: _____

_____ [1 mark]

2. Describe **three** ways in which alveoli are adapted for efficient gas exchange.

1 _____ [1 mark]

2 _____ [1 mark]

3 _____ [1 mark]

3. Emphysema is a health condition in which the alveoli walls break down, so the small air sacs join together to form larger air sacs.

A person with emphysema will get out of breath if they try to exercise.

Suggest an explanation of why. [3 marks]

Worked Example Emphysema happens in smokers and people working with asbestos. The shapes of their alveoli change and they can't breathe properly any more and need an oxygen tank.

This answer gains 0 marks.

The question is asking why the difference in the alveoli of a person with emphysema causes difficulty in getting all the oxygen they need to exercise. The student has concentrated on the symptoms of the disease rather than on what the changes in the alveoli imply. The student would have gained full marks for explaining that there would be less surface area for gas exchange to occur, and this would lead to less oxygen being absorbed into the blood. With less oxygen in the blood, there would be less oxygen available for respiration in the muscle cells and therefore less energy available for the muscles to contract.

Coronary heart disease

1. The table shows deaths due to coronary heart disease (2016) in England and Wales.

Age group	Deaths occurring in 2016	
	Males	Females
0–64	5750	1518
65–74	7072	2571
75–84	9954	5818
85+	8723	10117

Source: Office for National Statistics

a Which group had the **most** deaths from coronary heart disease in 2016?

_____ [1 mark]

Maths b What is the male-to-female ratio of deaths from coronary heart disease in the 0–64 age group? Give your answer to 2 significant figures.

Answer = _____ [2 marks]

2. Explain what happens to the heart in coronary heart disease.

_____

_____

_____ [3 marks]

3. Statins and stents are used to reduce the risk of coronary heart disease. Explain how they work.

Statins: _____

_____ [2 marks]

Stents: _____

_____ [2 marks]

4. Read the following article.

> If one or more of the valves in a heart are diseased or damaged, this can affect how blood flows through the heart. This can affect the supply of oxygen to the body's tissues. Replacing such valves will improve symptoms and quality of life. In valve replacement operations:
> - an incision is made in the middle of the breastbone
> - a heart–lung machine is used to circulate blood around the body during the operation
> - the heart is opened to replace the affected valve.
>
> Some patients may suffer from endocarditis, where the inner lining of the heart becomes infected. People having heart valve replacement surgery may also need to take blood-thinning medicine to prevent blood clots developing on the valve's surface.
>
> *Source: British Heart Foundation*

Evaluate the **advantages** and **disadvantages** of having a heart valve replacement operation. Use information from the article as well as your own knowledge.

Literacy

As this is an evaluation question, you are expected to include both advantages and disadvantages in your answers. If you focus on just one aspect, you will **not** gain full marks. Sometimes in questions like this you are also asked to give your own conclusion at the end.

_____

_____

_____

_____

_____

_____ [4 marks]

Risk factors for non-infectious diseases

1. The table shows some causes of adult deaths in Scotland.

Causes of adult deaths	Number of adult deaths (per 100 000 population)	
	Males	Females
Cancer	385	274
Coronary heart disease	165	105
Chronic obstructive pulmonary disease	71	58

Source: Office for National Statistics

Calculate the percentage of female deaths due to chronic obstructive pulmonary disease. Give your answer in standard form.

_____

_____

Percentage = _____% [1 mark]

2. Ali is at risk of developing coronary heart disease due to genetic factors. Suggest **three** lifestyle choices he could make to reduce the risk of developing coronary heart disease.

1 _____ [1 mark]

2 _____ [1 mark]

3 _____ [1 mark]

3. Describe the difference between a 'risk factor' for a disease and a 'causal mechanism' for the same disease.

_____

_____

_____ [2 marks]

4. Body Mass Index (BMI) is a person's mass (in kg) divided by the square of their height (in m).

The table shows the relationship between BMI and the percentage probability of developing Type 2 diabetes.

BMI	% probability of developing Type 2 diabetes	
	Males	Females
35.0+	70	74
30.0–34.9	56	54
25.0–29.9	30	34
18.5–24.9	20	17
<18.5	8	12

Maths **a** Describe **one** trend shown in the table.

_____

_____ [1 mark]

b Suggest an explanation for the trend described in part **a**.

_____

_____

_____ [2 marks]

Cancer

1. Which of the following are risk factors for cancer? Tick **three** boxes.

☐ Exposure to UV light

☐ Lack of exercise

☐ Lack of sleep

☐ Old age

☐ Smoking

[2 marks]

2. The table shows the detection rates and death rates for breast and prostate cancer in England in four different years.

The rates are per 100 000 people.

	2006	2011	2015	2016
Breast cancer death rate	41.5	37.0	34.3	34.1
Prostate cancer death rate	52.7	50.2	49.5	47.5
Breast cancer detection rate	162.3	164.1	170.5	167.9
Prostate cancer detection rate	172.9	177.3	179.4	173.7

Source: Office for National Statistics

Maths **a** Describe the trends in the detection rate and the death rate for prostate cancer.

_____

_____

_____

_____

[3 marks]

> **Remember**
>
> If a question asks you to describe trends or patterns in data, you should refer to changes in the data such as increases, decreases or changes in the rate of change. You need to describe trends over time and should not just describe differences in the data at specific points.

b Suggest explanations for the trends that you have described in part **a**.

_____

_____

_____

_____ [2 marks]

3. Suggest how a woman with breast cancer could also then develop liver cancer. [3 marks]

Worked Example

Cancer cells in the breast spread to different parts of the body through the blood. These cells can become secondary tumours in the liver or other organs.

> This response gains 2 marks. The student has correctly identified the spread of cancer to other parts of the body through the blood. Although they have mentioned that cancer cells can become secondary tumours, this is the same marking point as the idea of the cancer spreading. They need to describe the breast cancer as 'malignant' to gain the final mark.

Leaves as plant organs

1. The diagram shows a section through a leaf.

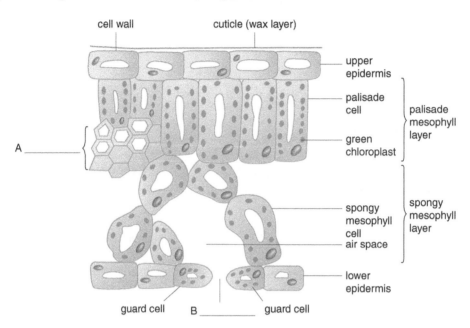

cell wall   cuticle (wax layer)

upper epidermis

palisade cell

palisade mesophyll layer

green chloroplast

A

spongy mesophyll cell

air space

spongy mesophyll layer

lower epidermis

guard cell   B   guard cell

a   Label parts **A** and **B** on the diagram.   [2 marks]

b   State the functions of parts **A** and **B**.

Part **A**: _____ [1 mark]

Part **B**: _____ [1 mark]

2. Explain how the following tissues are adapted for their functions.

a   Spongy mesophyll layer:

_____

_____ [2 marks]

b   Upper epidermal layer:

_____

_____ [2 marks]

3. Describe and explain how the structure of the palisade mesophyll layer is adapted for efficient photosynthesis.

_____

_____

_____

_____

_____ [4 marks]

Transpiration

1. Stomata are pores in the leaf that control gas exchange and water loss.

a Which cells control the size of the stomata?

_____ [1 mark]

b Describe the appearance of the stomata in the leaf of a well-watered plant during the daytime.

_____ [1 mark]

c Describe and explain the appearance of the same stomata if the plant is **not** watered for a few days. [2 marks]

Worked Example

They will close, otherwise the plant will die.

This answer gains 1 mark for stating that the stomata will close. If they didn't, the plant would wilt or droop from lack of water, but it is highly unlikely that the plant would die in a few days.

2. Describe and explain the process of transpiration.

_____

_____

_____

_____

_____

_____

_____

_____ [6 marks]

Translocation

. . . .

1. Translocation occurs through the phloem.

a Describe what is meant by the term 'translocation'.

_____

_____

_____ [2 marks]

b Describe the structure of phloem.

_____

_____

_____ [2 marks]

2. Most of the living tissue in a tree trunk is just under the bark.

If a tree is damaged and a ring of bark all around the trunk is removed, eventually the tree will die. Suggest an explanation for this. [3 marks]

The living tissue contains the phloem. So, without the phloem, food from the leaves cannot reach the roots, so they will die. This will eventually kill the tree.

3. Compare the movement of mineral ions and dissolved sugars in plants.

_____

_____

_____

_____

_____

_____

_____

_____

[6 marks]

Microorganisms and disease

1. What is meant by the term 'communicable' disease?

_____

_____ [1 mark]

2. Which of the following are communicable diseases? Tick **two** boxes.

☐ Cancer

☐ Diabetes

☐ Heart disease

☐ Malaria

☐ Measles [2 marks]

3. Explain what is meant by a 'pathogen'.

_____

_____ [1 mark]

4. Give **three** ways a pathogen can spread.

1 _____ [1 mark]

2 _____ [1 mark]

3 _____ [1 mark]

5. Compare the action of bacteria and viruses in causing infectious diseases. [6 marks]

Worked Example

Bacteria have cell walls and slime capsules to live outside the body. Once they get in, they rapidly reproduce, causing fever. Antibiotics kill bacteria. Viruses attack cells of the body and live inside them. They reproduce quickly and kill the cell when they burst out. Antibiotics and vaccines cannot kill viruses because they live in cells.

This is an extended response question and requires students to link ideas to produce a coherent and structured comparison of the spread of infectious diseases by bacteria and viruses.

The student made a number of relevant points: both bacteria and viruses reproduce rapidly in the body; bacteria produce symptoms such as fever; viruses live/reproduce inside cells; viruses cause cell damage; bacteria can be killed by antibiotics. However, there were also some errors: not all bacteria cause fever; the reason antibiotics don't kill viruses is because antibiotics disrupt cell processes and viruses aren't made of cells (although it is true that living inside cells makes it more difficult for our immune system to destroy viruses); vaccines can protect against many viruses. This answer gains 3 marks. The other 3 marks were lost because of the errors and because there were no actual comparisons made.

Viral diseases

1. Tobacco mosaic virus (TMV) is a plant pathogen.

a Describe **one** symptom of TMV.

_____

_____ [1 mark]

Worked Example

b Explain why this symptom affects the growth of the plant. [2 marks]

The virus makes the plant produce less chlorophyll and this means less photosynthesis.

This answer is only credited with 1 mark for the idea that less photosynthesis takes place. To gain the full 2 marks, it is important to explain why less chlorophyll affects the growth of plants by referring to the products of photosynthesis and how these are used for plant growth.

2. Compare measles and HIV.

_____

_____

_____

_____

_____

_____

_____ [4 marks]

3. The graph shows the number of laboratory confirmed cases of measles in England from 1996 to 2017.

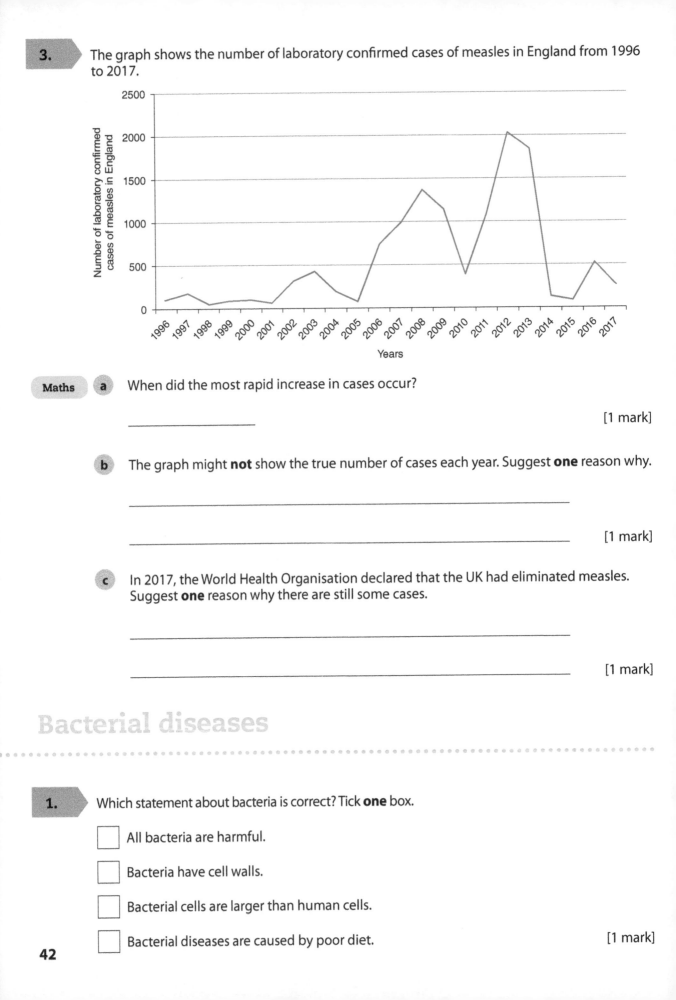

Number of laboratory confirmed cases of measles in England

Years

Maths **a** When did the most rapid increase in cases occur?

_____ [1 mark]

b The graph might **not** show the true number of cases each year. Suggest **one** reason why.

_____

_____ [1 mark]

c In 2017, the World Health Organisation declared that the UK had eliminated measles. Suggest **one** reason why there are still some cases.

_____

_____ [1 mark]

Bacterial diseases

. . . .

1. Which statement about bacteria is correct? Tick **one** box.

☐ All bacteria are harmful.

☐ Bacteria have cell walls.

☐ Bacterial cells are larger than human cells.

☐ Bacterial diseases are caused by poor diet. [1 mark]

2. The table shows the number of people in the UK diagnosed with gonorrhoea from 2013 to 2017.

Year	Number of people diagnosed with gonorrhoea in thousands	
	Female	Male
2013	8.5	22.7
2014	9.3	27.9
2015	9.2	32.2
2016	9.2	27.4
2017	11.5	33.1

Source: Public Health England

Maths **a** Compare the changes in the data for males and females with gonorrhoea from 2013 to 2017.

_____

_____

_____

_____ [3 marks]

Maths **b** What is the percentage increase of males with gonorrhoea from 2013 to 2017? Give your answer to 2 significant figures.

_____

_____

Percentage increase = _____ % [2 marks]

c Suggest **two** reasons for the increase in part **b**.

_____

_____

_____ [2 marks]

Command words

Remember that the command word 'suggest' means you aren't meant to have learnt the answer; instead you should use the knowledge you have to make sensible suggestions.

3. Compare *Salmonella* and gonorrhoea infections.

_____

_____

_____

_____

_____   [4 marks]

Malaria

1. Emma is bitten by an infected mosquito.

What type of organism is transmitted to her that causes malaria?

_____   [1 mark]

2. Which word best describes the role of mosquitoes in spreading malaria?

Tick **one** box.

☐ Carriers   ☐ Pathogens   ☐ Predators   ☐ Vectors   [1 mark]

3. Give **one** common symptom of malaria.

_____   [1 mark]

4. Describe and explain in detail **two** different ways the spread of malaria can be controlled.

_____

_____

_____

_____   [4 marks]

Exam technique

For questions like this one, try to give two answers that are quite different. Malaria can be controlled either by preventing mosquitoes from breeding or by preventing them from biting humans. So, in your answer try to give an example of a method for each way. Otherwise, if your two examples are too similar, you might not gain full marks.

5. UNICEF estimates that 5×10^8 people are infected with malaria every year. Of these people, 1 million die from malaria each year.

Maths What percentage of people infected with malaria die from the disease?

_____

_____

Percentage of infected people who die = _____ %   [2 marks]

Human defence systems

1. Describe how the following defend the body against pathogens.

a Skin:

_____

_____   [2 marks]

b Trachea and bronchi:

_____

_____

_____   [2 marks]

c Stomach:

_____

_____   [2 marks]

2. Some illnesses, such as AIDS, prevent the immune system from functioning normally.

Explain what would be the consequence of this.

_____

_____   [2 marks]

3. Describe the different ways in which white blood cells protect us from pathogens and their effects.

_____

_____

_____

_____

_____

_____

_____

_____ [6 marks]

Literacy
You should use correct terminology: for example, 'pathogens' **not** 'germs' or 'phagocytosis' instead of 'eating' pathogens.

Common misconception
Antibodies produced in immune responses do **not** remain in the body for a long period after the infection has ended. The correct idea is that white blood cells will have gained the ability to produce those antibodies quickly in response to a later infection by the same pathogens.

Vaccination

1. What does a vaccine contain?

_____ [1 mark]

2. Explain how a vaccine can help a person become immune to a disease.

_____

_____

_____

_____

_____ [4 marks]

3. Explain why a vaccine is **not** effective as a cure for people who already have the disease.

_____

_____ [1 mark]

4. Unlike vaccines for most diseases, new vaccines for flu are needed each year.

Suggest an explanation for why new flu vaccines are needed so frequently.

_____

_____

_____

_____ [3 marks]

5. The graph shows the antibody level in blood after a vaccination (**A**) and after a second exposure to the same pathogen (**B**).

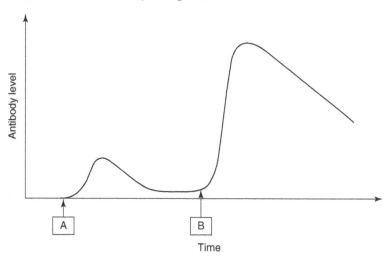

Explain why the increase in the antibody level after **A** is different from the increase in antibody level after **B**. [3 marks]

This response gains 1 mark. The student has been let down by weak expression of ideas. First of all, the suggestion that 'measles' is injected is wrong. It is the 'measles pathogen/virus' which is injected. Measles is a description of a set of symptoms and not, in itself, a virus. Secondly, white blood cells do not 'remember' a response but they produce memory cells that are able to respond faster to subsequent infections, and produce antibodies more quickly and in greater quantities.

 After measles is injected into the person, the antibody levels rise because the measles made white blood cells produce antibodies. When the measles entered the body again (**B**) the white blood cells were very quick to make lots of antibodies because they remembered the infection.

Antibiotics and painkillers

1. Draw lines from each type of drug to the illnesses they will treat.

	Gonorrhoea
Antibiotic	Headache
	HIV
	Malaria
Painkiller	Measles
	Toothache

[3 marks]

2. Painkillers treat the symptoms but **not** the causes of disease.

Explain what this statement means.

_____

_____ [1 mark]

3. Explain why it is difficult to develop drugs to treat diseases caused by viruses.

_____

_____

_____ [2 marks]

4.

a If you go to your doctor with a chest infection, they may **not** prescribe antibiotics.

Explain why.

_____

_____

_____

_____

_____

_____ [4 marks]

b Suggest under what circumstances a doctor would prescribe antibiotics.

Remember
Antibiotic-resistant bacteria are becoming more common. There are questions about this in the 'Evidence for evolution' section.

_____

_____

_____ [2 marks]

Making and testing new drugs

1. Complete the sentences.

The heart drug digitalis was originally found in _____. [1 mark]

The antibiotic penicillin was discovered by _____. [1 mark]

The painkiller aspirin was originally found in _____. [1 mark]

2. Scientists wanted to compare the effectiveness of two painkillers, drug **A** and drug **B**.

They chose 100 volunteers who were suffering pain.

They gave half the volunteers a dose of drug **A** and the other half a dose of drug **B**.

The volunteers recorded how much pain they felt in the next 24 hours.

a Is the test valid? Explain your answer.

_____

_____

_____ [2 marks]

b Suggest **two** factors that should be matched in the volunteers.

1 _____ [1 mark]

2 _____ [1 mark]

c Suggest why the scientists used 100 volunteers.

_____

_____ [1 mark]

d Before testing on volunteers, the drugs had been tested on animals. Suggest why.

_____ [1 mark]

3. In the trial of a drug for reducing the risk of heart disease, some patients were given a placebo drug.

a What is in a placebo drug?

_____

_____ [1 mark]

b Why are placebos used?

_____

_____ [1 mark]

c This trial was a double-blind trial. What is meant by a 'double-blind' trial?

_____

_____ [1 mark]

d   Why are double-blind trials used?

_____

_____   [1 mark]

Monoclonal antibodies

1.   The diagram shows a cancer cell being treated using monoclonal antibodies.

Higher Tier only

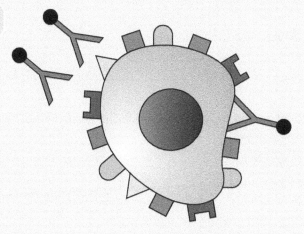

a   Label a monoclonal antibody on the diagram.   [1 mark]

b   What are monoclonal antibodies?

_____   [1 mark]

2.   The diagram shows some stages in the production of monoclonal antibodies.

Higher Tier only

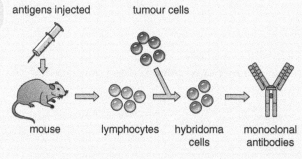

antigens injected   tumour cells

mouse   lymphocytes   hybridoma   monoclonal
                          cells   antibodies

a Explain why antigens are injected into the mouse.

_____ [1 mark]

b Tumour cells are combined with lymphocytes to make hybridomas.

Explain why.

_____

_____

_____ [2 marks]

c Explain why the antibodies produced are monoclonal.

_____ [1 mark]

3. **a** Explain how and why monoclonal antibodies can be used in the treatment of cancer.

Higher
Tier only

_____

_____

_____

_____

_____ [4 marks]

b State **two** other uses of monoclonal antibodies.

1 _____ [1 mark]

2 _____ [1 mark]

c Explain why monoclonal antibodies are **not** as widely used as scientists had first hoped.

_____

_____ [1 mark]

Plant diseases

1. Gardeners should wash **any** tools that have been used on plants infected with disease.

Suggest why.

_____ [1 mark]

2. Rose bushes can suffer from a disease called black spot. This causes black or purple spots on the surface of leaves, which often turn yellow and drop early.

a What causes black spot?

_____ [1 mark]

b Explain why the black spot pathogen affects the growth of the infected plant.

_____

_____

_____ [2 marks]

c If left untreated, black spot can quickly infect all rose bushes in a garden.

State **two** ways by which the black spot pathogen naturally spreads.

1 _____ [1 mark]

2 _____ [1 mark]

d Describe how to treat rose black spot and prevent it spreading. [3 marks]

Worked Example

Make sure you remove all the affected leaves and stems immediately and burn them. If you don't do this, spores from the infected plant can easily be spread to other plants. Another way is to use fungicides to get rid of the infection.

This answer is worth 3 marks because the student has described three points about the immediate removal of stems and leaves, the need to burn them, and also mentions fungicides. Another point that could have been made is that infected parts of the plant should **not** be composted as spores can survive and re-infect other rose plants.

3.

Higher Tier only

Testing kits using monoclonal antibodies can be used to detect and identify plant diseases. However, plants do **not** produce antibodies themselves.

Suggest how monoclonal antibodies could be obtained in order to identify a plant disease.

_____

_____

_____ [2 marks]

Identification of plant diseases

1.

Higher Tier only

A gardener notices spots on the leaves of her tomato plants.

State **two** methods that could be used to identify the disease.

1 _____ [1 mark]

2 _____ [1 mark]

2.

Higher Tier only

Describe **four** observations used to detect plant diseases.

1 _____ [1 mark]

2 _____ [1 mark]

3 _____ [1 mark]

4 _____ [1 mark]

3.

Plant growth is affected if the plants are deficient of certain mineral ions.

For each plant below, identify the mineral ion deficiency and explain the reason for the appearance of the plant.

a Plant **A** has pale leaves with yellow patches. It is **not** growing well.

Deficiency: _____ [1 mark]

Explanation of appearance:

_____

_____ [1 mark]

b Plant **B** shows very stunted growth, even though it gets lots of light and is well watered.

Deficiency: _____ [1 mark]

Explanation of appearance:

_____

_____ [1 mark]

4. Aphids are an example of an insect pest that can damage crop plants.

Suggest how aphids reduce crop yield.

_____

_____

_____ [2 marks]

Plant defence responses

1. Draw **three** lines to match the plant descriptions to the type of defence response shown.

Hawthorn trees have thorns on their branches.	Physical defence against microorganisms.
Nettle leaves have stinging hairs.	Chemical defence.
Basil stems have layers of dead cells.	Mechanical defence against animals.

[2 marks]

2. Plants such as peppermint and sage contain oils. Suggest how these oils form part of their defence responses.

_____

_____

_____

_____ [3 marks]

3. Acacia trees are found in warm tropical areas. They grow quickly and reach heights of up to 12 m. Their branches have sharp spines, 8–10 cm long. These spines provide shelter for a species of stinging ant. Acacia leaves and branches contain a chemical called tannin, which is distasteful to animals.

Explain how the adaptations in acacia trees help protect them.

_____

_____

_____

_____

Hint

Examination questions sometimes refer to examples that students may **not** be aware of. Do not worry about these as there will always be enough information given in the question to allow you to answer.

_____ [4 marks]

4. The leaves of the mimosa plant curl inwards and droop if touched.

Suggest explanations for how this response could be a defence response against herbivores. [2 marks]

If the leaves curl this could make it more difficult for herbivores to eat the leaves. Also, the animals may think that the plant is infected with disease and so avoid eating its leaves. If it happens quickly it might also scare off the animal.

This is a very good answer and would gain full marks. In this question it is not made clear how many explanations are required. The student has looked at the mark allocation and given not only two ideas, but also a third in case one of the others wasn't mark-worthy.

Photosynthesis reaction

1. Green plants make their own food through photosynthesis.

Most photosynthesis occurs in the leaves.

Complete each sentence below.

a The light energy needed for photosynthesis is absorbed by a chemical called

_____.   [1 mark]

b Since energy is absorbed during photosynthesis, this means it is an

_____ reaction.   [1 mark]

Synoptic **c** The gases used and produced in photosynthesis enter and leave a leaf through pores

called _____.   [1 mark]

2. Write a **word** equation for photosynthesis.

_____   [2 marks]

3. Write a **symbol** equation for photosynthesis.

_____   [3 marks]

> **Hint**
> Chemical equations should each be written on a single line, with an arrow between the reactants and products. Don't forget that symbol equations should be balanced.

4. There are sub-cellular structures in many plant cells which carry out photosynthesis.

Synoptic **a** Name these structures.

_____   [1 mark]

b   Three types of cell in a leaf contain the structures in part **a**.

Name these **three** types of cell.

Draw a circle around the name of the type of cell that contains **most** of the structures.

1 _____

2 _____

3 _____   [4 marks]

5.   Scientists estimate there is a total of 7×10^{12} tonnes of carbon dioxide in the atmosphere and that photosynthesis removes 10×10^{10} tonnes of this each year.

Maths   Calculate the percentage of the total carbon dioxide in the atmosphere that is used by photosynthetic organisms each year.

Give your answer to 2 significant figures.

_____

_____

Answer = _____ %   [2 marks]

Rate of photosynthesis

1.   Tomato growers can alter the conditions in greenhouses to make tomato plants grow faster.

What conditions will make tomato plants grow faster? Tick **two** boxes.

☐ Increasing the temperature.

☐ Increasing the oxygen concentration in the air.

☐ Increasing the nitrogen concentration in the air.

☐ Turning the lights on at night.   [2 marks]

2.   Explain what is meant by the 'rate of photosynthesis'.

_____

_____   [1 mark]

3.

Required practical

The diagram shows the apparatus that Laura used to measure the rate of photosynthesis under different light intensities.

A lamp was placed at different distances from the beaker containing pondweed. The investigation was repeated several times and the average (mean) number of oxygen bubbles produced per minute at each distance was calculated.

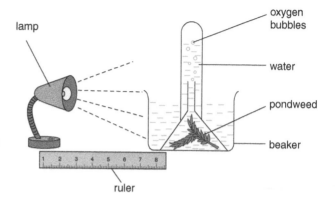

a State **two** factors which must be kept constant during the investigation.

1 _____ [1 mark]

2 _____ [1 mark]

b What is the dependent variable in this experiment?

_____ [1 mark]

This is a table of Laura's results.

Distance of lamp from pondweed in cm	Average (mean) number of bubbles produced per minute
10	106
15	48
20	29
25	16
30	12
35	8

Maths **c** Describe and explain what Laura can conclude from these results.

_____

_____

_____ [2 marks]

Maths   **d**   Do the results obey the inverse square law for photosynthesis? Explain your answer using data from the table.

Higher Tier only

_____

_____

_____

_____ [2 marks]

4.   Lee investigated how the colour of light affects the rate of photosynthesis.

This is a table of his results:

Colour of light bulb	Amount of oxygen produced in 5 minutes in cm³			
	1st measurement	2nd measurement	3rd measurement	Mean
Red	24	19	21	21
Green	6	4	3	4
Blue	32	34	32	

Maths   **a**   Calculate the mean number of bubbles produced in the blue light to complete the table. [1 mark]

b   Lee's classroom has a fish tank with a built-in light.

The light encourages the growth of green algae on the sides of the tank.

What colour bulb would be best to **reduce** the growth of the algae?

Explain your answer.

_____

_____

_____ [3 marks]

Limiting factors

1. The graph shows the effect of the concentration of carbon dioxide on the rate of photosynthesis in lettuce plants at 20 °C.

Maths **a** Which letter shows the highest rate of photosynthesis?

Choose **A**, **B** or **C**.

_____ [1 mark]

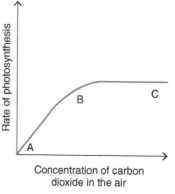

b What is the limiting factor for photosynthesis at point **A**?

Explain your answer.

Limiting factor: _____ [1 mark]

Explanation: _____

_____ [1 mark]

c Suggest what is the limiting factor for photosynthesis at point **C**.

Explain your answer.

Limiting factor: _____

Explanation: _____

_____ [2 marks]

2. The graph shows the effect of light intensity on the rate of photosynthesis at two different temperatures.

Higher Tier only

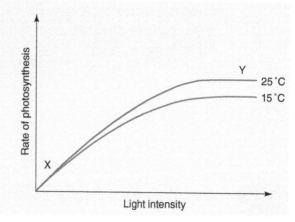

a Describe the trends shown in the graph.

_____

_____

_____ [2 marks]

b Explain why the graphs are the same at point **X** but different at point **Y**.

_____

_____

[2 marks]

_____

c A farmer grows tomatoes in a glasshouse.

He adds extra lights inside the glasshouse to increase the rate of photosynthesis.

He does this to grow more tomatoes in a year and make more profit.

Suggest and explain **two** possible reasons why he might **not** make more profit. [4 marks]

Worked Example ▶ If he adds extra lights he might increase the rate of photosynthesis but only if there are no other limiting factors. He might find he also needs to increase the temperature, for example. The extra lighting, and possibly heating, will cost money, so although he might grow and sell more tomatoes, he might not actually make any extra profit.

This is a very good answer and would gain full marks. The student has given two suggestions and explained them. Another possible limiting factor could be the concentration of carbon dioxide, and it could be that the farmer would need to increase the carbon dioxide concentration before increasing light intensity would have any effect.

Uses of glucose from photosynthesis

1. Complete the following sentences.

Synoptic The energy needed for photosynthesis comes from _____. [1 mark]

Energy is absorbed by a green pigment called _____. [1 mark]

If the temperature is decreased, the rate of photosynthesis will _____. [1 mark]

2. Some of the glucose made in photosynthesis is converted into starch for storage.

a Explain why starch is used for storage. [2 marks]

Worked Example

Starch is insoluble, which means that it can be easily stored.

This answer would gain 1 mark for stating that starch is insoluble. There is no mark for the second part of the answer as it only repeats what is in the question. The second mark is for explaining the importance of starch being insoluble. Sugars are soluble so can be transported around the plant in solution (in the phloem tissue). However, conversion to insoluble starch means that it will stay in the storage tissues.

b Give **three** other ways in which plants use glucose made in photosynthesis.

Hint

Be specific about the molecules that glucose is converted to. Do **not** make vague references to growth, repair or reproduction without specific details of how glucose could contribute to these processes.

1 _____ [1 mark]

2 _____ [1 mark]

3 _____ [1 mark]

3. The roots of plants take up nitrates from the soil.

a Explain why nitrates are needed by plants.

_____

_____ [2 marks]

Synoptic **b** Describe the appearance of a plant deficient in nitrates.

_____

_____ [2 marks]

Cell respiration

....

1. Organisms release energy in respiration.

Give **two** reasons why organisms need energy.

Exam technique

Do **not** say something vague such as 'to stay alive'. Give a specific reason why energy is needed.

1 _____ [1 mark]

2 _____ [1 mark]

2. Give **one** way in which aerobic respiration is different from anaerobic respiration.

_____ [1 mark]

3. Write a **word** equation for aerobic respiration.

_____ [2 marks]

4. Write a **symbol** equation for aerobic respiration.

_____ [3 marks]

5. Ali and James investigated the effect of temperature on the rate of aerobic respiration in locusts. They recorded the rate of respiration over 5 minutes at two different temperatures.

The table shows their results.

Time (mins)	Rate of respiration in arbitrary units	
	10 °C	20 °C
5	2.5	6.0

Maths **a** The locusts took in 54 mm³ of oxygen during the 5 minutes at 20 °C. Calculate the average (mean) volume of oxygen the locusts took in each minute.

_____

Volume of oxygen taken in = _____ mm³ per minute [1 mark]

b Suggest **one** reason why the rate of respiration was lower at 10 °C.

_____

_____ [1 mark]

c Apart from temperature, suggest **two** other factors that would affect the rate of aerobic respiration.

1 _____ [1 mark]

2 _____ [1 mark]

d Ali suggested they should do the experiment several times at each temperature.

Explain why this would improve the investigation. [2 marks]

> This answer gained 0 marks as it is not specific enough. Doing the experiment several times would make the results more 'repeatable' and would allow any anomalous results to be identified.

To make sure the results are correct.

6. Explain the term 'metabolism', using examples in your answer.

_____

_____

_____

_____

_____ [4 marks]

Anaerobic respiration

1. The equation shows anaerobic respiration in muscles.

glucose → lactic acid

Give **three** ways this is different from **aerobic** respiration in muscles.

1 _____

2 _____

3 _____ [3 marks]

2. The diagram shows an experiment to investigate anaerobic respiration in yeast cells.

a What gas is collected in the gas syringe?

_____ [1 mark]

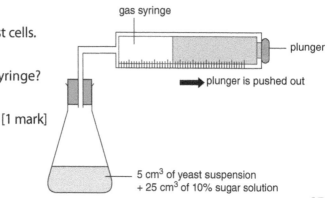

gas syringe

plunger

plunger is pushed out

5 cm³ of yeast suspension + 25 cm³ of 10% sugar solution

b Describe how you could use the apparatus above to measure the rate of reaction in the conical flask.

_____

_____ [1 mark]

3. **a** Write a word equation for anaerobic respiration in yeast.

_____ [1 mark]

b Give **two** examples where this type of respiration is important in the food and drink industry.

1 _____ [1 mark]

2 _____ [1 mark]

4. Compare anaerobic respiration in yeast cells with anaerobic respiration in muscle cells. [4 marks]

Command words

The command word 'compare' requires a description of the similarities and differences. Make sure you do **not** just write about one type of anaerobic respiration.

Anaerobic respiration takes place in muscle cells and uses glucose to produce energy and lactic acid, whereas anaerobic respiration in yeast cells produces ethanol.

This answer gains 1 mark out of a possible 4. Although the student is correct in what they've written, they have **not** given as much detail as they could have. They have correctly identified one difference, i.e. the production of lactic acid in muscle cells compared with ethanol in yeast cells, but they should also have included the other difference, that carbon dioxide is produced in yeast cells but not in muscle cells. They should also have included the similarities that both processes release energy from glucose without the use of oxygen.

Response to exercise

1. The table shows how much energy is used by two women of different masses running at different speeds.

Running speed in km/h	Energy used in kJ/h	
	48 kg woman	72 kg woman
6.5	1205	1807
12.0	2310	3464

Maths **a** When the 48 kg woman runs at 12.0 km/h instead of 6.5 km/h, how much more energy does she use each hour?

_____

Answer = _____ kJ/h   [1 mark]

Maths **b** Express the answer to part **a** as a percentage increase.

Give your answer to 3 significant figures.

_____

Answer = _____ %   [1 mark]

c The energy consumption for the 72 kg woman is higher than for the 48 kg woman. Suggest why.

_____

_____   [1 mark]

2. When a person starts to run, the muscles need more oxygen.

Give **three** ways in which the body reacts to the increased demand for oxygen.

1 _____   [1 mark]

2 _____   [1 mark]

3 _____   [1 mark]

3.

Explain what is meant by 'oxygen debt'.

In your answer include how the body responds to an oxygen debt.

_____

_____

_____

_____

_____

_____

_____

[6 marks]

Homeostasis

· ·

1. Which of the following are controlled by homeostasis?

Tick **two** boxes.

☐ Body mass

☐ Body temperature

☐ Lung capacity

☐ Reaction time

☐ Water content of body   [2 marks]

2. Homeostasis is maintained by control systems.

Draw lines to match each part of a control system to its function and then to examples.

Part of control system	**Function**	**Examples**
Receptor	Processes information	Brain and spinal cord
Coordination centre	Detects changes in the environment	Muscles and glands
Effector	Brings about a response	Eyes and skin

[3 marks]

3. Explain the importance of homeostasis in the human body.

_____

_____

_____   [2 marks]

4. What are the **two** types of control system in the human body?

1 _____   [1 mark]

2 _____   [1 mark]

The nervous system and reflexes

1. The diagram shows the pathway for a simple reflex action.

sharp pin — Neurone A

Neurone C

gap

Neurone B

muscle — spinal cord

a What type of neurone is neurone **A**?

_____ [1 mark]

b What is the name for the gap between neurones?

_____

[1 mark]

c How does information pass across the gap?

_____ [1 mark]

d How can you tell that this reflex action is **not** a conscious action?

Use information from the diagram to help you answer.

_____ [1 mark]

e Describe what happens to the muscle when it receives an impulse from neurone **C**.

_____ [1 mark]

f Reflex actions like this one are useful. Explain why.

_____

_____ [2 marks]

Scientific terms
Wherever possible, always use the correct scientific terms. For example, in question part **e**, words such as 'tightens' or 'tenses' are **not** acceptable.

2. Chloe and Ben investigated the effect of caffeine on reaction time.

Required practical

- Chloe held a ruler just above Ben's hand.
- Chloe let go of the ruler. Ben caught it as quickly as possible.
- They repeated this experiment five more times.
- Ben then drank a soft drink containing caffeine.
- 20 minutes after the drink, Chloe and Ben repeated the first three steps.

The table shows their results.

Distance ruler fell before it was caught in cm	
Before drinking caffeine	**After drinking caffeine**
21	10
18	11
21	8
19	14
23	13
15	10
Mean =	Mean = 11

Maths
An answer should **not** be given to more significant figures than shown by those used in the calculation.

Maths **a** Calculate the mean result for 'before drinking caffeine'.

Write your answer in the table. Give your answer to the correct number of significant figures. [2 marks]

Maths **b** What do the results show about the effect of caffeine on reaction time?

_____

_____ [1 mark]

c Give **one** reason why your conclusion in part **a** may **not** be valid.

_____

_____ [1 mark]

d Suggest **two** ways in which they could improve their investigation.

1 _____

_____ [1 mark]

2 _____

_____ [1 mark]

3. Describe the differences between receptors and effectors.

_____

_____

_____

_____ [4 marks]

The brain

1. The diagram shows a human brain.

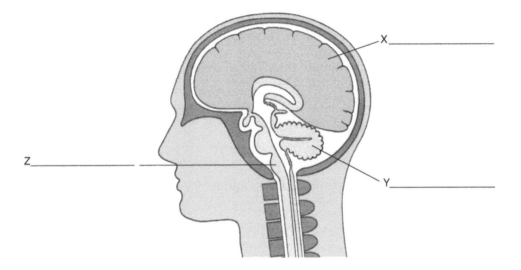

a Label the **three** parts **X**, **Y** and **Z**. [3 marks]

b Give **one** function of part **X**.

_____ [1 mark]

c Give **one** function of part **Y**.

_____ [1 mark]

d Give **one** function of part **Z**.

_____ [1 mark]

2. Jo has a head injury.

Symptoms include:

- loss of balance
- problems with coordination of limbs.

a Suggest which part of her brain has been damaged.

_____ [1 mark]

Higher
Tier only **b** Give **one** technique doctors could use to help identify which part of the brain has been damaged.

_____ [1 mark]

Higher
Tier only **c** Suggest why this technique might **not** give a conclusive answer. [2 marks]

*Worked
Example* The brain is very complicated and different parts may be involved in coordination and balance.

> This answer gains 1 mark. The student could have given more detail explaining that most functions are **not** localised in one area.

The eye

· ·

1. The diagram shows the eye.

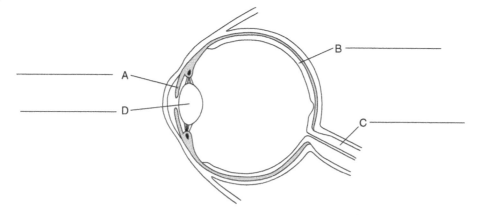

a Label parts **A**, **B**, **C** and **D** on the diagram. [4 marks]

b Draw a line to the part that provides strength and structure to the eye. Label this part **E**. [1 mark]

2. Draw lines to show which part of the eye carries out each function.

Function	Part
Carries nerve impulses to brain	Cornea
Protects eye	Optic nerve
Refracts light	Retina
Senses light	Sclera

[3 marks]

3. Describe the changes in the pupil and iris when going **from** a brightly lit environment **to** a dimly lit environment.

_____

_____

_____

_____ [3 marks]

4. A person is looking at a nearby object and then looks at another object further away in the distance. Describe and explain the changes in the eye when it focuses light from the distant object.

_____

_____

_____

_____

_____ [4 marks]

Remember
The ciliary muscles and suspensory ligaments supporting the lens change the shape of the lens to focus light from objects at different distances – this is called **accommodation**. Do not confuse this with the radial and circular muscles in the iris, which change the size of the pupil according to how much light is available – this is called **adaptation**.

Seeing in focus

1. A person stops looking at a distant object and starts looking at a nearby object. How do the lenses in their eyes change?

Underline the correct answer.

become fatter   **become thinner**   **stay the same**   [1 mark]

2. **a** As they grow older, some people cannot clearly see nearby objects, although they can clearly see more distant objects.

What is this condition called?

_____   [1 mark]

b Suggest an explanation why this condition occurs.

_____

_____   [2 marks]

c This condition can be corrected by wearing glasses.

Write down the type of lens used in these glasses.

_____   [1 mark]

3. The diagram shows an eye of a person with short-sight (myopia) looking at a distant object.

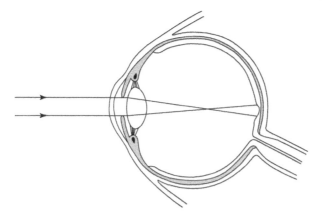

Draw a lens in front of the eye that can correct short-sightedness.

Draw rays of light on the diagram to show how the new lens corrects short-sight.   [3 marks]

Wearing glasses is one way to correct defective eyesight.

Give **three** other ways to correct defective
eyesight. [3 marks]

Hard contact lenses, soft contact lenses or
surgery to change the shape of the cornea.

Although the student has given
three answers, in this case they
would only get 2 marks, as the
mark scheme only had 1 mark for
reference to contact lenses. If you
get a question like this, try to give
quite different examples, rather than
different versions of the same idea.
The student could have got a third
mark for lens replacement surgery.

Control of body temperature

1. A scientist measured the volume of sweat produced by a person in one afternoon.

The table shows the results.

Time (pm)	Volume of sweat in cm³
12.00–12.59	20
1.00–1.59	30
2.00–2.59	600
3.00–3.59	880
4.00–4.59	800
5.00–5.59	25

a Suggest what happened at 2.00 pm. Explain your answer.

_____

_____ [2 marks]

Maths b What percentage of the total sweat produced between 12.00 pm and 5.59 pm was
produced between 2.00 pm and 4.59 pm? Give your answer to 2 significant figures.

_____

_____

Answer = _____ % [2 marks]

**Higher
Tier only** c How does sweat help to control body temperature?

_____

_____ [2 marks]

2.

Higher Tier only

It has been snowing. Joe goes outside without a coat. Describe and explain the changes that occur in his body when the temperature of his surroundings decreases.

_____

_____

_____

_____

_____

_____

_____ [6 marks]

Common misconceptions

The blood vessels that constrict or dilate to regulate temperature are small arteries (called arterioles) – capillaries **cannot** constrict or dilate. Also, no blood vessels 'move' to be nearer to, or further from, the surface of the skin.

Hint

Read the question carefully! It asks about temperature regulation in the *human* body; therefore, discussion of hairs erecting ('goose bumps') and trapping air is not relevant.

Hormones and the endocrine system

1.

The diagram shows some of the endocrine (hormone) glands in the human body.

Label glands **A**, **B**, **C** and **D** on the diagram.

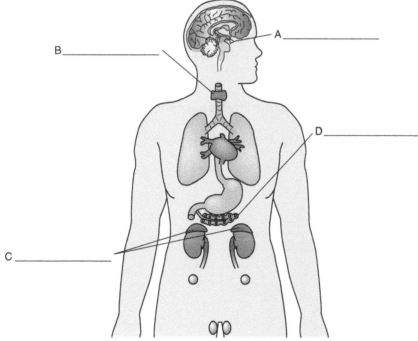

[4 marks]

2. **a** Which gland is known as the 'master gland'?

_____ [1 mark]

b Explain why it is known as the 'master gland'.

_____

_____ [1 mark]

3. Describe the differences between nervous and hormonal control.

Synoptic

_____

_____

_____

_____

_____

_____

_____ [6 marks]

> **Remember**
> If you are describing differences, do **not** refer, for example, just to nervous control. Instead, refer to **both** nervous and hormonal control to explain both sides of each difference.

Controlling blood glucose

1. In people with diabetes the blood glucose level is **not** controlled properly.

Describe each type of diabetes and how it is normally treated.

a Type 1 diabetes:

Description: _____ [1 mark]

Treatment: _____ [1 mark]

b Type 2 diabetes:

Description: _____ [1 mark]

Treatment: _____ [1 mark]

2. The table shows some information about diabetes in the UK.

Number of people diagnosed with diabetes in 2017	3.7×10^6
Percentage of people diagnosed with diabetes who have Type 2 diabetes	90

Maths **a** Calculate the number of the people in the UK in 2017 diagnosed with Type 2 diabetes.

_____

_____

Number of people with Type 2 diabetes = _____ [1 mark]

b The percentage of people diagnosed with Type 2 diabetes is increasing.

Suggest **one** reason why.

_____

_____ [1 mark]

3.

Higher Tier only

Worked Example

Describe the negative feedback process that controls the blood glucose level in someone without diabetes. [6 marks]

Remember
Do **not** confuse glycogen with glucagon.

If glucose levels rise after a meal then the pancreas releases insulin, which causes glucose to move from the blood and converts it to glycogen, which is stored in the liver and muscles. If glucose levels fall the pancreas releases glucagon, which converts glycogen back into glucose, which goes back into the blood.

This is a good answer which covers the main points, gaining 4 out of a possible 6 marks. However, it misses out some details about the negative feedback process which works to keep the blood glucose level **constant**. Insulin only causes **excess** glucose to be removed from the blood, and glucagon only converts enough stored glycogen back into glucose to bring the blood glucose level back to normal.

Maintaining water and nitrogen balance in the body

1. One way water is lost from the body is in urine.

State **two** other ways water is lost from the body.

1 _____ [1 mark]

2 _____ [1 mark]

2. **a** Proteins are digested to smaller molecules.

Synoptic What is the name of these smaller molecules?

_____ [1 mark]

Higher Tier only **b** If there is an excess of these smaller molecules they are converted to another substance and excreted.

Name the substance that is excreted.

_____ [1 mark]

3. Describe the processes that form urine. [4 marks]

Worked Example The kidneys filter the blood, removing urea, ions, water and glucose. The substances that are needed in the body are reabsorbed back into the blood and the substances that are not needed make up the urine.

> The student gains 2 marks for describing how filtration by the kidneys will remove most of the contents of the blood, and for naming the substances involved. However, the second part of the answer is too vague. The student should have said that there is 'selective reabsorption' of the useful substances (all the glucose, some ions and some water); and that excess water, excess ions and all the urea are **not** reabsorbed but instead are removed from the body as urine.

4. Explain how ADH controls the amount of water excreted by the kidneys.

Higher Tier only

_____

_____

_____

_____

_____

_____

_____

[6 marks]

Hormones in human reproduction

1. These are some of the events occurring during the menstrual cycle.

A	ovulation occurs
B	oestrogen is secreted by cells around the maturing egg
C	FSH is secreted by the pituitary gland
D	oestrogen stimulates growth of uterus lining
E	an egg starts to mature in an ovary

Put the stages in the **correct** sequence. The first one has been done for you.

C				

[2 marks]

2. The diagram shows the levels of three hormones during pregnancy. The baby is usually born at about 40 weeks.

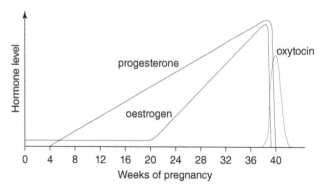

a Compare how the levels of oestrogen and progesterone change from 0 to 38 weeks.

_____

_____

_____

_____

_____ [4 marks]

> **Remember**
> 'Compare' means write about both things, giving **both** similarities and differences.

b One of the hormones shown on the graph causes the uterus to contract when the baby is born.

Suggest which hormone and explain the reason for your choice.

Hormone: _____ [1 mark]

Reason: _____ [1 mark]

3.

Higher Tier only

Describe the interactions between FSH, LH and oestrogen during the menstrual cycle.

_____

_____

_____

_____ [3 marks]

Contraception

1. Read the information about two types of contraception used by women.

Combined pill:	Contraceptive implant:
• contains two hormones	• contains one hormone
• 99% effective at preventing pregnancy	• placed under skin of upper arm by doctor
• increases chance of headaches	• 99% effective at preventing pregnancy
• slightly increases chance of breast cancer	• lasts for 3 years
• increases risk of blood clots.	• reduces heavy or painful periods
	• periods become irregular.

a) Which **two** hormones are used in contraceptive pills?

1 _____ [1 mark]

2 _____ [1 mark]

b) Suggest **two** advantages of using an implant rather than taking contraceptive pills.

1 _____ [1 mark]

2 _____ [1 mark]

c) Suggest **two** disadvantages of using an implant rather than taking contraceptive pills.

1 _____ [1 mark]

2 _____ [1 mark]

2. Describe the barrier methods of contraception used in men and women.

_____

_____

_____

_____

_____ [4 marks]

Using hormones to treat human infertility

1. A woman has **not** been able to become pregnant. Her doctor recommends In Vitro Fertilisation (IVF) treatment.

Higher Tier only In the first stage of IVF the woman is injected with the hormones FSH and LH.

a) Explain why the woman is injected with FSH and LH.

_____ [1 mark]

b Eggs from the woman are then fertilised with sperm.

Where does this happen?

_____ [1 mark]

The fertilised eggs develop into embryos.
Some of the embryos are inserted into the woman's uterus.

c Why are several embryos inserted and **not** just one?

_____ [1 mark]

d Explain the risks of inserting several embryos and **not** just one.

_____

_____

_____ [3 marks]

e Suggest **two** other disadvantages of this kind of fertility treatment.

1 _____ [1 mark]

2 _____ [1 mark]

f The IVF treatment described above is **not** suitable for treating all forms of infertility.

Suggest **one** cause of infertility that could **not** be treated in this way.

_____ [1 mark]

Negative feedback

• •

1. Explain what is meant by 'negative feedback'.

Higher
Tier only

_____

_____ [1 mark]

2.

Higher Tier only

Synoptic

Worked Example

Using sweating as an example, explain the role of negative feedback in the control of body temperature. [4 marks]

If your body gets too hot then you produce sweat, which evaporates to cool you down. When the brain detects that your body temperature is back to normal, then sweating will stop; otherwise you would lose too much heat and your body temperature would fall below its normal level.

> This is a good answer and gains 4 marks. The student has clearly described a negative feedback system that responds to reverse the effects of any change, in this case reducing body temperature after it has risen.

3.

Higher Tier only

a Where is adrenaline produced?

_____ [1 mark]

b Explain the role of adrenaline in the body.

_____

_____

_____

_____

_____ [4 marks]

> **Remember**
> It is **not** the presence of sweat on the skin that has a cooling effect but the evaporation of water in the sweat which transfers heat from the body.

4.

Higher Tier only

The thyroid gland produces the hormone thyroxine.

Thyroxine plays an important role in growth and development because it stimulates the basal metabolic rate.

a What is meant by 'basal metabolic rate'?

_____ [1 mark]

b Some people have a condition which means their thyroid does **not** release enough thyroxine.

Suggest **one** effect this might have.

_____ [1 mark]

Plant hormones

1.

Required practical

Mia placed germinating bean seeds in different positions in a Petri dish containing moist cotton wool and left it in a dark cupboard. The diagram shows the results after several days.

a   Describe and explain her results.

_____

_____   [2 marks]

b   Name this type of growth response.

_____   [1 mark]

c   Suggest **one** reason why this type of growth response is useful for young seedlings.

_____   [1 mark]

d   Why did Mia place the dish in a dark cupboard?

_____   [1 mark]

2.

Required practical

Liam investigated the effect of light, shining from one side, on the growth of plant shoots.

The diagram shows a shoot at the start and after two days.

a   Name this type of growth response.

_____   [1 mark]

b   Suggest **one** reason why this type of growth response is useful for young plants.

_____   [1 mark]

c This type of growth response is controlled by a plant hormone called auxin. Explain how auxin causes this response.

_____

_____ [2 marks]

3.

Higher Tier only

a Name the plant hormone involved in starting seed germination.

_____ [1 mark]

Higher Tier only

b Name the plant hormone involved in fruit ripening.

_____ [1 mark]

Uses of plant hormones

1.

Higher Tier only

Draw lines to match each use to the plant hormone used.

You may use each hormone once, more than once, or not at all.

Use	plant hormone

| As a weed killer | Auxin |

| To promote flowering | Ethene |

| To help plant cuttings grow roots | Gibberellin |

[3 marks]

2.

Higher Tier only

Some scientists investigated the effect of ethene concentration on the ripening of bananas. The table shows their results.

Ethene concentration as percentage of air	0	1	2	3	4	5
Percentage of bananas that were ripe after three days	19	31	58	82	92	92

Maths

a Draw a graph to show ethene concentration against percentage of bananas that were ripe after three days. Draw a line of best fit. [3 marks]

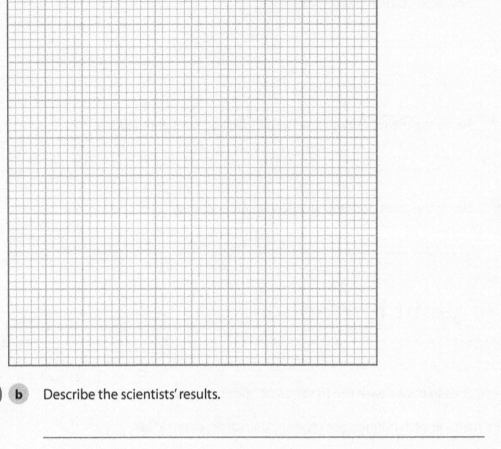

Maths **b** Describe the scientists' results.

_____

_____

_____ [2 marks]

c Suggest **one** control variable the scientists used.

_____ [1 mark]

Bananas are transported before they ripen to avoid damage.

Sometimes ethene gas is used to ripen them before they are sold.

If this does **not** happen, bananas will eventually ripen naturally as they produce their own ethene.

d Ripening bananas should be kept away from other fruit.

Suggest an explanation of this.

_____

_____ [2 marks]

Sexual and asexual reproduction

1. Strawberry plants produce shoots called runners that grow sideways along the ground.

New strawberry plants grow from the runners.

The new plants have the same inherited characteristics as the original plant.

Exam technique
Carefully read through the sentences with your chosen word to check that they make sense in terms of language, as well as science. If a sentence does **not** make grammatical sense, it is unlikely that the correct word has been chosen.

a Complete the sentences using words from the box.

asexual	differentiation	embryos	fertilisation	gametes
genes	meiosis	mitosis	sexual	

The new strawberry plants are produced by _____ reproduction.

In this type of reproduction, body cells divide by _____ .

The new plant has the same _____ as the parent plant.   [3 marks]

b The new plants are clones. Explain why.

_____

_____

_____   [2 marks]

c The new plants will **not** all grow to exactly the same height.

Suggest why.

_____

_____   [1 mark]

2. A mule is produced when a male donkey and a female horse breed together.

Remember
Sexual reproduction does **not** have to involve sexual intercourse (think of flowers for example), so do **not** use that in your answer.

a A mule is produced by sexual reproduction.

Explain what is meant by sexual reproduction.

_____

_____ [1 mark]

b A mule has characteristics of both a donkey and a horse.

Explain why.

_____

_____

_____ [2 marks]

3. Complete the table to name the different types of gamete.

Gametes	Flowering plants	Animals
Female	egg cells	
Male		

[3 marks]

4. Some organisms can reproduce both sexually and asexually, for example strawberry plants, daffodils and many fungi.

Explain the advantages and disadvantages of sexual and asexual reproduction. [6 marks]

This answer gained 4 marks. The student has clearly discussed the advantage of genetic variation in allowing a species to survive in a changing environment. There is also an understanding that sexual reproduction requires energy. Further marks could have been gained if the student had mentioned that asexual reproduction is usually faster than sexual reproduction and can produce many identical offspring when conditions are favourable.

Worked Example

Asexual reproduction is well suited for organisms that remain in one place and are unable to find mates. However, asexual reproduction does not lead to variation between organisms, meaning that entire groups can be wiped out by disease, or if the environment changes.

Sexual reproduction produces variation in the offspring. Therefore, species that reproduce sexually can adapt to new environments and are less likely be wiped out by a single disease. However, sexual reproduction requires energy on the part of the organism to find a mate. It is not well suited to organisms that are isolated or stuck in one place.

Cell division by meiosis

1. A chimpanzee has 48 chromosomes in each body cell.

a How many chromosomes are in a chimpanzee egg cell?

_____ [1 mark]

b How many chromosomes are in a **fertilised** chimpanzee egg cell?

_____ [1 mark]

c The cells in a fertilised chimpanzee egg cell divide to form an embryo.

How many chromosomes are in each cell of the embryo?

_____ [1 mark]

2. Put the stages of meiosis in the correct order by writing **1**, **2** and **3** in the table.

Stage	Order
Cell division forms four cells	
Cell division forms two cells	
The DNA in each chromosome is copied	

[1 mark]

3. Describe **one** similarity and **one** difference between the gametes produced by the same individual.

Similarity: _____

_____ [1 mark]

Difference: _____

_____ [1 mark]

4. Compare the processes of cell division by meiosis and mitosis. [6 marks]

Synoptic

Worked Example

In mitosis (used for growth, repair and asexual reproduction) there is one cell division that produces two daughter cells that contain the full number of chromosomes and are genetically identical. In meiosis (used for sexual reproduction) there are two cell divisions which produce four daughter cells, each with the half number of chromosomes. They are genetically different from each other.

> This answer gains 5 marks. The student has set out a logical and coherent answer, writing about both meiosis and mitosis correctly and pointing out characteristics of the cells produced. The student could have gained full marks by pointing out where both processes took place: in humans, meiosis takes place in the ovaries and testes, and mitosis in all other body cells.

DNA, genes and the genome

1. Write numbers in the boxes to put the following in order of size, starting with **1** for the smallest.

☐ cell ☐ chromosome ☐ gene ☐ nucleus [1 mark]

2. Explain **one** difference between the terms 'DNA' and 'gene'.

_____

_____ [1 mark]

> **Hint**
> If you're asked to describe or explain the difference between two things, make sure you refer to **both** in your answer.

3. Genes control many of our characteristics, such as our eye colour.

Synoptic Explain briefly how they do this.

_____

_____

_____ [3 marks]

4. Explain what is meant by the term 'genome'.

_____ [1 mark]

5. Give **three** reasons why it is important to study the human genome.

1 _____ [1 mark]

2 _____ [1 mark]

3 _____ [1 mark]

Structure of DNA

1. The diagram shows a short section of DNA.

a Name parts **A** and **B**.

A _____ [1 mark] ——————— part A

B _____ [1 mark] ——————— part B

bases

b Part **X** is made up of one base, one part **A** and one part **B**.

What is the name of part **X**?

_____ [1 mark]

c DNA is a polymer.

What is meant by the term 'polymer'?

_____ [1 mark]

d The whole DNA molecule is a double helix.

What does 'double helix' mean?

_____

_____ [2 marks]

2. The diagram shows the bases on one strand of a section of DNA.

A T G T A C C T A

Higher Tier only

a Write out the sequence of bases on the complementary strand.

_____ [2 marks]

b How many amino acids are coded for by the strand of DNA?

_____ [1 mark]

c What is the name of a section of DNA which codes for a protein?

_____ [1 mark]

3.

In a sample of DNA, 32% of the bases are T. Complete the table to show the percentages of the **other three** bases in the sample.

Base	Percentage (%)
A	
T	32
C	
G	

Hint
Remember how the bases are paired.

[3 marks]

Protein synthesis and mutations

1.

a Write numbers in the boxes to put the stages of protein synthesis in the correct order.

Stage	Order
The completed protein chain leaves the ribosome and folds into a unique shape.	
The mRNA copy leaves the nucleus.	
Amino acids are joined together in the correct sequence at the ribosome.	
An mRNA copy is made of a section of DNA.	
Carrier tRNA molecules bring amino acids to the ribosome in the correct order.	
The mRNA joins with a ribosome.	

[2 marks]

b Using **one** example, explain why the shape of a completed protein is important.

_____

_____ [2 marks]

2. Read the article.

> ### Sickle cell
>
> Sickle cell is an inherited disorder caused by a mutation in the gene coding for haemoglobin.
>
> Haemoglobin is the protein in red blood cells that carries oxygen.
>
> Cells containing sickle cell haemoglobin can become deformed in shape and unable to carry oxygen efficiently around the body.
>
> Sickle cell is caused by a single base change in the DNA coding for haemoglobin.

Explain why the mutation in the haemoglobin gene affects the ability to carry oxygen around the body.

_____

_____

_____

_____

_____ [4 marks]

3. Some DNA is non-coding.

a What does 'non-coding' mean?

_____ [1 mark]

b Give **one** function of non-coding DNA.

_____ [1 mark]

Inherited characteristics

1. The ability to taste a chemical called PTC is controlled by a single gene with two alleles. Being able to taste PTC is caused by the dominant allele **T**. The recessive, non-tasting allele is **t**.

a Explain what is meant by the term 'allele'.

_____ [1 mark]

b Explain the difference between 'dominant' and 'recessive' alleles.

_____

_____

_____ [2 marks]

A couple who can both taste PTC have two children. One child can taste PTC and one cannot.

c What is the **phenotype** of the two parents?

_____ [1 mark]

d What are the **genotypes** of the two parents? Explain the reason for your answer.

_____

_____ [2 marks]

Maths **e** Complete the Punnett square to show the possible genotypes of the children.

Mother's gametes

Father's gametes

[2 marks]

Maths **f** According to this Punnett square, what is the probability of the couple having a child who is **not** able to taste PTC? Explain your answer.

_____

_____

_____ [2 marks]

Maths **g** Does this probability match the actual children of this couple? Explain the reason for this.

_____

_____ [1 mark]

2. In mice, black fur is caused by a dominant allele **B**. The recessive allele **b** codes for brown fur.

a What is the genotype of a mouse with brown fur?

_____ [1 mark]

Maths **b** A black mouse could be either homozygous or heterozygous.

Higher Tier only Suggest how you could use breeding experiments to work out whether a particular mouse with black fur is homozygous or heterozygous.

Include Punnett square diagrams in your answer.

[4 marks]

_____

_____

_____

_____

Inherited disorders

1. Polydactyly is an inherited condition causing extra fingers or toes. It is caused by a dominant allele **D**.

Tom and his mother have polydactyly, but his father does **not**. Neither does Tom's sister.

a Draw a family tree to show Tom's family, using the key provided.

> **Key**
>
> Male with polydactyly   ■
>
> Male without polydactyly   □
>
> Female with polydactyly   ●
>
> Female without polydactyly   ○

[2 marks]

b What are the genotypes of Tom, his parents and sister?

Tom: _____

Mother: _____

Father: _____

Sister: _____

> **Hint**
> Look at the family tree to help you work out the possible genotypes.

[4 marks]

2. Cystic fibrosis is an inherited disorder caused by a recessive allele **f**.

Ali has the genotype **ff**. His wife, Shara, has the genotype **FF**. They want to have children but worry about their children or grandchildren having cystic fibrosis. They ask their GP about this.

Suggest what the GP would tell them. [4 marks]

Worked Example

Cystic fibrosis is caused by a recessive allele, so to have the condition you would have to be ff. All the children will be **Ff** so there is no possibility that any will have cystic fibrosis.

> This answer gains 2 marks for clearly explaining why the children could **not** have cystic fibrosis. However, the question also asked about grandchildren and the student should also have addressed this. Any children would be carriers, so there is a possibility that any grandchildren could have the condition if their other parent was **ff** (had the condition) or **Ff** (was a carrier).

Sex chromosomes

1. Choose the correct words, letters or numbers from the box to complete the sentences.

Synoptic

| XX fertilisation 23 XY 46 meiosis YY 92 mitosis |

The sex chromosomes in the human male are _____.

Most human body cells contain _____ chromosomes.

The gametes (sex cells) each contain _____ chromosomes.

Gametes are produced by _____ . [4 marks]

2. **a** Name the part of the sperm cell that contains the chromosomes.

Synoptic

_____ [1 mark]

Maths **b** A man releases 300 million sperm cells at a time. How many of these sperm cells contain an X chromosome?

_____ [1 mark]

3. **a** Complete the Punnett square to show how sex is inherited.

Maths

Man

Woman X		
X		

[1 mark]

Maths **b** What is the probability of a couple's first child being a girl?

_____ [1 mark]

Maths **c** A couple have one girl. What is the probability of the couple's next child also being a girl?

_____ [1 mark]

Variation

1. Put ticks in the table to show how each characteristic is controlled.

Characteristic	Controlled by		
	Genes only	Environment only	Both genes and environment
Blood group			
Eye colour			
Height			
Language spoken			
Scar			
Skin colour			

[3 marks]

2. When a puppy grows up it will look similar to its parents but will **not** be identical to either one of them.

a Explain why it will look similar to its parents.

_____

_____ [1 mark]

Synoptic **b** Suggest **two** reasons why it will **not** be identical to either of its parents.

1 _____ [1 mark]

2 _____ [1 mark]

3. Explain what a mutation is and how it might affect an organism's phenotype. [6 marks]

Synoptic

Higher Tier only

Worked Example

Mutations change the sequence of bases in DNA. The sequence of amino acids coded for by the gene therefore can be changed and this will change the protein produced. This will affect the phenotype if the protein is significantly altered. However, sometimes the change to the DNA has little or no effect on the protein or the phenotype.

This answer gains 4 marks. The student has made a good attempt at explaining what happens when mutations affect the coding sequences of DNA. The student has also recognised that mutations can have a small or large effect on the phenotype. However, there was no attempt to explain about mutations in non-coding sequences.

Theory of evolution

1. Complete the following sentence.

The theory of evolution states that life began on Earth _____

years ago and that _____ life forms have evolved into

_____ life forms. [3 marks]

2. Giraffes have long necks, which help them feed on leaves from tall trees.

This means they can eat food that is **not** available to other species.

a The ancestors of modern giraffes did **not** have such long necks.

Use Darwin's theory of natural selection to explain how giraffes evolved their long necks.

_____

_____

_____

_____

_____ [4 marks]

b Lamarck suggested a different theory of evolution.

How would Lamarck have explained how giraffes
evolved their long necks? [2 marks]

This answer gains full marks for a clear explanation of Lamarck's theory.

Worked Example

Giraffes spent their lives reaching for food in tall trees. This stretched their necks and the long necks were passed on to their offspring. Therefore, in each generation the necks got longer.

Synoptic **c** Suggest why modern scientists would **not** agree with Lamarck's explanation.

_____

_____

_____ [2 marks]

3. It took many years for Darwin's theory of evolution by natural selection to become widely accepted.

Suggest why.

_____

_____

_____

_____

_____ [3 marks]

Darwin and Wallace

1. Charles Darwin and Alfred Russel Wallace developed a theory of evolution.

What is the name of this theory?

_____ [1 mark]

2. The term 'survival of the fittest' is often used to describe Darwin and Wallace's theory of evolution.

Suggest what this term means.

_____

_____ [1 mark]

3. Darwin made many observations supporting his theory on a long sea voyage around South America in the 1830s. However, he did not publish his theory until over 20 years later in 1859.

a   What was the name of the book he published in 1859?

_____ [1 mark]

b Suggest why Darwin waited so long before he published his ideas.

_____

_____

_____ [2 marks]

4. Wallace suggested a similar theory after making observations in Southeast Asia.

One observation was that many species use warning colours to deter predators.

Explain how warning colours could evolve.

_____

_____

_____

_____ [3 marks]

Speciation

1. An Alsatian and a Border Collie are two different breeds of dog but they are the **same** species.

An Alsatian and a wolf are described as **different** species even though sometimes they do breed together to produce a puppy.

Use these examples to explain the meaning of the term 'species'.

_____

_____

_____ [2 marks]

Alsatian

Border Collie

Wolf

2. Lake Victoria is a very large lake in East Africa. It contains a type of fish called a cichlid. Two species of cichlid are shown.

There are over 500 different species of cichlid in Lake Victoria. Over tens of thousands of years, the water level in the lake has risen and fallen many times. When the water level falls, separate lakes form. When the water level rises again, the smaller lakes join together to form one large lake.

Suggest how the changes in water level have given rise to so many species of cichlid.

_____

_____

_____

_____

_____

_____

_____

_____

_____ [6 marks]

3. Scientists who study fossils often find it difficult to decide whether similar fossil skeletons of different extinct animals belong to the same species or are from different species.

Synoptic Suggest why they have this difficulty.

_____

_____

_____

_____ [3 marks]

The understanding of genetics

1. Our understanding of how genes are inherited began with the work of Gregor Mendel.

Which of the following describes Mendel's work?

Put ticks in the boxes to show your answers.

☐ Mendel carried out breeding experiments on plants.

☐ Mendel used a microscope to study genes.

☐ Mendel was the first person to discover genes, although he called them 'units'.

☐ Mendel worked out the structure of DNA. [2 marks]

2. Put the discoveries below in the order that they were made. Show your answer by putting numbers in the boxes.

☐ How chromosomes behave during cell division.

☐ That chromosomes and 'units' behave in similar ways.

☐ That inheritance is determined by 'units' that are passed on to descendants unchanged.

☐ The structure of genes and DNA. [2 marks]

3. Mendel crossed pure-bred tall plants with pure-bred short plants.

The offspring in the first generation all grew into tall plants.

He then bred the first generation plants with each other to produce a second generation. The result was a mixture of tall plants and short plants.

> **Remember**
> Pure-bred means that when he bred them together he only ever got the same type.

When he repeated this, he found the second generation always contained **three times** as many tall plants as short plants.

The diagram shows his results.

Parents: pure-bred tall plants × pure-bred short plants

First generation: all tall plants

Second generation: tall plants + short plants

Ratio: 3 : 1

We now know that there are **two alleles** controlling height.

T is the dominant allele for being tall.

t is the recessive allele for being short.

Synoptic
Maths

a   Complete the Punnett squares to explain Mendel's results.

Crossing the parents:

	Pure-bred tall	
	T	T
Pure-bred short		

Crossing the first generation:

	Tall	
Tall		

[4 marks]

b   Mendel bred pea plants because he could get hundreds of offspring. He considered using mice for his breeding experiments.

Suggest why plants gave him better results.

_____

_____

_____

[2 marks]

Hint

The use of the command word 'suggest' tells you that you are **not** expected to have learnt this answer, nor be an expert on pea plants or mice. Use the information in the question, as well as your own knowledge, to come up with sensible ideas.

4.   Mendel did his work in the mid-nineteenth century, but it was **not** until the twentieth century that its importance was recognised.

Suggest why this took so long.

_____

_____

_____   [2 marks]

Evidence for evolution

1. Which of the following provide evidence for evolution?

Put ticks in the boxes to show your answers.

☐ Climate change

☐ Development of antibiotic resistance in bacteria

☐ Extinction of species

☐ Fossils

☐ Similarities between living species [3 marks]

2. The diagram shows an evolutionary tree for humans and other apes.

gibbons   orangutans   gorillas   chimpanzees   bonobos humans

Time

a Which **two** species are most closely related?

Explain your answer.

_____

_____ [2 marks]

b Is it correct to say that humans have evolved from chimpanzees? Explain your answer. [2 marks]

No, it is not correct. Humans and chimpanzees do share a relatively recent common ancestor but that ancestor was neither a chimpanzee nor a human, but some other kind of ape.

In an exam you won't get a mark for simply saying yes or no. The marks here are for the explanation. This answer gains full marks.

Common misconception

Most species alive today have **not** evolved from another species that is also alive today. If two species are very similar, then what we can usually say is that they have evolved from a relatively recent common ancestor, which was also a similar species.

3. **a** Explain what fossils are and the different ways they can form.

_____

_____

_____

_____

_____

_____ [4 marks]

b Give **two** reasons why there are many 'gaps' in the fossil record.

1 _____ [1 mark]

2 _____ [1 mark]

4. **a** Explain how antibiotic resistance evolves in bacteria.

_____

_____

_____

_____

_____

_____

_____ [6 marks]

b Give **two** ways to reduce the rate of development of antibiotic-resistant bacteria.

1 _____ [1 mark]

2 _____ [1 mark]

Extinction

1. The black rhinoceros lives in Africa. It has several subspecies (varieties). One subspecies, the western black rhinoceros, was officially declared extinct in 2011, as none had been seen since 2006. The other surviving subspecies are 'endangered' and in serious danger of extinction.

a Explain the difference between the terms 'endangered' and 'extinct'.

_____

_____

_____ [2 marks]

b Suggest **one** reason why the western black rhinoceros became extinct.

Hint
You do **not** need to have studied the black rhinoceros to answer the questions. Use what you have studied and apply it to this example.

_____

_____ [1 mark]

c Suggest **three** ways to prevent the surviving black rhinoceros subspecies becoming extinct.

1 _____ [1 mark]

2 _____ [1 mark]

3 _____ [1 mark]

Selective breeding

1. Domesticated dogs are descended from wild wolves.

All modern breeds of dog are the result of selective breeding over thousands of years.

Suggest **one** characteristic that the first dogs were selected for and explain why.

Characteristic: _____ [1 mark]

Explanation: _____ [1 mark]

2. Edible apples are descended from crab apples, which are small, hard and bitter.

Suggest **three** features that edible apples have been selectively bred for.

1 _____ [1 mark]

2 _____ [1 mark]

3 _____ [1 mark]

3. **a** Race horses have been selectively bred to run fast. Describe the process of selectively breeding race horses.

Hint
The process of selective breeding is the same regardless of the species concerned, so just apply what you know about selective breeding to this example.

_____

_____

_____

_____

_____ [5 marks]

b Describe the problems associated with selective breeding.

_____

_____

_____

_____ [3 marks]

Genetic engineering

1. Explain what is meant by the term 'genetic engineering'.

Remember
'Genetic engineering', 'genetic modification' and 'GM' all mean the same thing.

_____

_____ [1 mark]

2. Explain the advantage of genetic engineering compared with selective breeding.

_____

_____ [2 marks]

3. Give **three** ways plant crops can be improved by genetic engineering.

1 _____ [1 mark]

2 _____ [1 mark]

3 _____ [1 mark]

4. People with Type 1 diabetes need to take regular injections of insulin. In the past the insulin was extracted from animals such as cows and pigs. Now, most insulin comes from bacteria that have been genetically engineered to contain the gene for human insulin.

a Suggest advantages of obtaining insulin from genetically-engineered bacteria rather than from animals. [3 marks]

Worked Example

The genetically-engineered bacteria produce insulin made using a human gene so it is identical to insulin made normally in the human body. Insulin from animals may be similar to human insulin but it is not identical and some people may have an allergic reaction to it.

The student has made some valid points and this answer gains 2 marks. The student could also have explained that insulin can be obtained in much larger quantities from bacteria because, given the right conditions, they can reproduce very rapidly producing millions of bacterial cells, all making insulin. Another valid point is that some people might object to using insulin from animals, for example, vegetarians.

Higher Tier only **b** Describe the process of genetically engineering bacteria to produce human insulin.

_____

_____

_____

_____

_____

_____

_____ [6 marks]

Cloning

1. One method of cloning is 'embryo transfer'. This is sometimes used when selectively breeding cattle.

   a   Write numbers in the empty boxes to show the order of the stages in embryo transfer.

Stage	Order
Collect sperm from selected bull and eggs from selected cow	
Cells of embryo split apart forming separate embryos	
Each embryo implanted into a surrogate cow	
Fertilisation using IVF	
Fertilised egg grows into an embryo	

[2 marks]

Synoptic   b   Explain the advantage of this method over traditional selective breeding.

_____

_____

_____   [2 marks]

2. A garden centre grows and sells rose bushes. One type is very popular because of the colour of its flowers. The garden centre wants to reproduce as many of this type of rose bush as it can.

   a   One method of reproduction is to let the plant be pollinated naturally and then grow new rose bushes from the seeds it produces.

   Explain **one** disadvantage of this method for the garden centre.

_____

_____   [2 marks]

b Two other methods of reproduction are tissue culture and taking cuttings. Describe the advantages and disadvantages of these two methods for the garden centre.

_____

_____

_____

_____

_____ [4 marks]

3. A zorse is a hybrid resulting from a cross between a male zebra and a female horse.

Zorses are unable to breed because they do **not** produce fertile eggs or sperm.

However, scientists could reproduce zorses using 'adult cell cloning'.

Describe how adult cell cloning could be used to clone an adult zorse.

_____

_____

_____

_____

_____

_____ [6 marks]

Classification of living organisms

1. **a** In the eighteenth century a Swedish scientist developed a classification system that is still used today.

What was his name?

_____ [1 mark]

b How did the Swedish scientist decide whether species should be in the same classification group?

_____ [1 mark]

c Put the classification groups in order from the largest to the smallest. The first one has been done for you.

Class	
Family	
Genus	
Kingdom	1
Order	
Phylum	
Species	

[2 marks]

d The Swedish scientist gave the human species its scientific name *Homo sapiens*.

What is this system of naming species called?

_____ [1 mark]

e Two extinct species closely related to modern humans are *Homo habilis* and *Australopithecus afarensis*.

Which one is more closely related to modern humans? Explain your answer.

_____

_____ [1 mark]

2.

In 1977, Carl Woese suggested that living organisms should be classified into a 'three-domain' system. One domain is the eukaryota, which consists of all eukaryotes. The other two domains consist of prokaryotes.

Synoptic   **a**   Name **two** kingdoms that are part of the eukaryota.

1 _____ [1 mark]

2 _____ [1 mark]

b   Name the **two** domains that consist of prokaryotes.

1 _____ [1 mark]

2 _____ [1 mark]

c   What technology did Woese use to construct the three domains that was **not** available to the Swedish scientist?

_____ [1 mark]

Habitats and ecosystems

1. A lake contains different species of fish. Some feed on water plants and others feed on smaller fish or insects. Herons hunt fish, and make their nests in the trees around the lake. Herons also hunt frogs, which eat insects living in and around the water. The insects feed on plants or other insects.

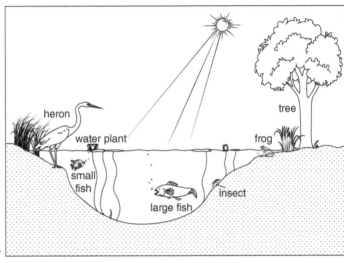

a Using the lake and the organisms living in and around it as examples, explain the meanings of the following terms: **population**, **community**, **habitat** and **ecosystem**.

_____

_____

_____

_____

_____

_____ [4 marks]

> **Hint**
> For each term make sure you give a definition and include an example.

b A disease kills all the frogs living in and around the pond.

Suggest and explain how this might affect the other organisms.

_____

_____

_____

_____

_____

_____

_____

_____ [6 marks]

2. Explain what is meant by a 'stable' community.

_____

_____ [1 mark]

Food in an ecosystem

1. The diagram shows part of an African food web.

Use the diagram to answer the questions.

a Name a **herbivore**.

_____ [1 mark]

b Name a **producer**.

_____ [1 mark]

c Name an organism that is both **predator** and **prey**.

_____ [1 mark]

d Explain which organism is the **apex predator**.

Apex predator: _____ [1 mark]

Explanation: _____ [1 mark]

e Name an organism in the third trophic level.

_____ [1 mark]

f Write a food chain with **four** trophic levels.

_____

_____ [2 marks]

g Explain how energy enters this food web.

_____

_____

_____ [4 marks]

Abiotic and biotic factors

1. Put ticks in the table to show whether each factor is biotic or abiotic.

Factor	Biotic	Abiotic
Disease		
Food availability		
Moisture level		
Predators		
Soil pH		
Temperature		

[3 marks]

2. Give **three** things that animals compete for.

1 _____ [1 mark]

2 _____ [1 mark]

3 _____ [1 mark]

3. Bluebells are small plants that live in woodland. They grow and produce flowers in early spring before the trees in the woodland have grown their leaves. Suggest an explanation for this.

_____

_____ [2 marks]

4. Lynx are predators living in northern Canada. They hunt snowshoe hares.

Maths The graph shows how the populations of these animals changed over many years.

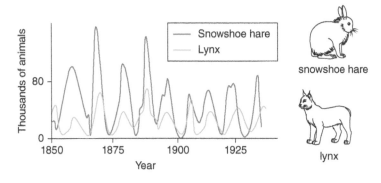

Describe and explain the patterns shown in the graph. [6 marks]

Common misconception

It is a common mistake when answering questions like this for students to write about animals 'dying out' when they mean that the numbers are decreasing. If the animals really did 'die out' there would be none left to continue the pattern.

The numbers of both animals rise and fall but the lynx numbers rise and fall just after the hare numbers. This is because the lynx feed on the hares. When there are lots of hares, there is lots of food for the lynx so they produce many young and their population rises. Eventually, however, there are so many lynx hunting the hares that the hare population falls. This means there is not enough food for all the lynx so their numbers fall. Now there are not many lynx hunting them, so the hare numbers rise and the whole cycle starts again.

This a very good answer and would gain 6 marks. The student has **described** the pattern and then gone on to **explain** it. Full marks have been gained for the explanation because it goes through the whole of the cycle.

Adapting for survival

1. Put ticks in the table to show whether each adaptation is structural, behavioural or functional.

Adaptation	Structural	Behavioural	Functional
Acacia trees have sharp thorns to deter herbivores from feeding on the leaves.			
Camels can tolerate higher body temperatures than most mammals to help them survive in the desert.			
Eagles have sharp talons for catching their prey.			
Swallows migrate in the summer from Africa to Europe in order to breed.			
Weaver birds build very complicated nests to lay their eggs in.			

[5 marks]

2. Some organisms, such as some bacteria, are described as 'extremophiles'.

Explain what this term means.

_____

_____

_____

[2 marks]

3. Water lilies have large leaves that float on water.

Synoptic

Water lily leaves have their stomata on their upper surface, unlike most plants, which usually have stomata on the lower surface of their leaves.

Suggest an explanation for the position of stomata in water lilies.

_____

_____

_____ [2 marks]

4.

Synoptic

Poison dart frogs are very poisonous. They also have bright colouration which makes them very conspicuous.

Suggest an explanation for these adaptations.

_____

_____

_____ [2 marks]

Measuring population size and species distribution

1.

Required practical

Hannah investigated how the presence of a tree in a park affected the growth of grass below it.

She marked out a line from the tree trunk to open grassland.

She placed a quadrat every metre and recorded the percentage of ground covered by grass.

The table shows her results.

Distance from tree trunk (m)	Percentage cover of grass (%)
0	0
1	24
2	67
3	88
4	100
5	99

a Hannah placed the quadrats in a line.

What is the name of this type of line?

_____ [1 mark]

b The diagram shows the quadrat that she used.

Explain how Hannah used this quadrat to work out the percentage cover of grass. [2 marks]

Place the quadrat on the ground. This type of quadrat has 100 small squares. Count the number of squares in which grass is growing. This gives you the percentage cover.

This answer gains 1 mark. The student has missed out one important point, which is that you only count a square if more than half the area is taken up by grass.

c Describe and explain Hannah's results.

_____

_____

_____

_____ [3 marks]

2. Liam used a quadrat to estimate the number of dandelion plants in the park.

He placed the quadrat in random positions and counted the number of dandelions in the quadrat each time.

The table shows his results.

Quadrat position	Number of dandelions
1st	3
2nd	0
3rd	0
4th	1
5th	0
6th	2
7th	0
8th	0
9th	0
10th	2

Maths **a** The size of the quadrat was 0.25 m² and the size of the park was 120 m².

Use this information and Liam's results to calculate an estimate of the number of dandelions in the park.

_____

_____

_____

Number of dandelions in park = _____ [3 marks]

b Why is it important that the quadrat was placed in random positions?

_____ [1 mark]

c How should Liam have made sure the quadrats were placed randomly?

_____ [1 mark]

_____ [1 mark]

d How could Liam have made his estimate more accurate?

_____ [1 mark]

Cycling materials

1. The diagram shows part of the carbon cycle.

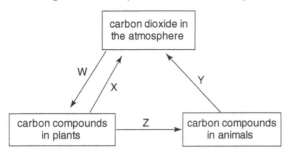

a Identify the processes labelled **W**, **X**, **Y** and **Z**.

W _____ [1 mark]

X _____ [1 mark]

Y _____ [1 mark]

Z _____ [1 mark]

> **Common misconception**
> Many people incorrectly think that plants take in carbon dioxide and store it as carbon dioxide gas, which later can be released. What really happens is that the carbon in carbon dioxide is used to build the materials that plants (and other living things) are made of, such as carbohydrates, lipids and proteins.

b Describe **two** other ways, **not** shown in the diagram, that carbon found in animals and plants can eventually return to the atmosphere.

1 _____

_____ [2 marks]

2 _____

_____ [2 marks]

2. Describe the processes involved in the water cycle.

_____

_____

_____

_____ [4 marks]

3. Explain why it is important for living organisms that materials, such as carbon and water, are recycled.

_____

_____

_____ [2 marks]

Decomposition

1. A class investigated the effect of temperature on the rate of decay of fresh milk by measuring pH change. Their method is shown below.

Required practical

1. Add 5 ml of milk, 7 ml of sodium carbonate solution, and 5 drops of phenolphthalein to a test tube.

2. Place the test tube in a water bath until contents all reach the same temperature.

3. Add 1 ml of lipase into the test tube and start a stop clock.

4. Stir the contents of the test tube until the solution loses its pink colour; record the time.

5. Repeat steps 1–4 for a range of different temperatures.

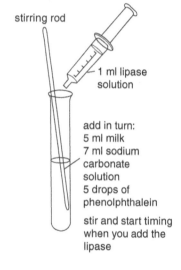

stirring rod

1 ml lipase solution

add in turn:
5 ml milk
7 ml sodium carbonate solution
5 drops of phenolphthalein

stir and start timing when you add the lipase

a Phenolphthalein is a pH indicator.

Explain why it is used in this experiment.

_____

_____ [2 marks]

The table shows one group's results.

Temperature in °C	Time taken for pink colour to disappear in s
5	545
15	264
25	105
35	34
45	27
55	98

Maths **b** At which temperature was the rate of reaction fastest?

_____ °C   [1 mark]

Maths **c** Calculate the fastest rate of reaction in s^{-1}.

Higher Tier only Give your answer in standard form.

_____   [1 mark]

Rate of reaction = _____ s^{-1} [1 mark]

Maths **d** Describe and explain the effect of temperature on the rate of reaction shown in the results.

Synoptic

_____

_____

_____

_____

_____   [4 marks]

2. Biogas generators decay waste material to produce methane gas to be used as a fuel.

a Suggest **one** waste material that can be used in biogas generators.

_____   [1 mark]

b What are the best conditions for decay inside a biogas generator?

_____

_____

_____ [3 marks]

c How would the conditions be different to encourage decay in a compost heap?

_____ [1 mark]

Changing the environment

1. Draw lines to match the species affected by environmental change with the type of change.

Higher
Tier only

Species affected by environmental change	Type of change

| Pigeons that live in cities are descended from wild doves that naturally live and nest on rocky cliffs. | Geographic |

| Pink-footed geese breed in Greenland in summer but spend the winter in Britain where it is warmer. | Human interaction |

| Surtsey Island formed near Iceland in 1963 due to volcanic activity. Now a range of species live there. | Seasonal |

[2 marks]

2. Polar bears hunt on Arctic sea ice, feeding mainly on seals. In the summer the ice melts and the bears come on land, where less food is available. Rising temperatures are causing sea ice to melt earlier in the spring and form later in the autumn. For every week earlier that the ice melts, the bears come on land with approximately 10 kg less body mass.

Higher
Tier only

a Explain why the body mass of polar bears when they come on land is decreasing.

_____

_____ [2 marks]

b Suggest how the distribution of polar bears might change as a result of the changes in sea ice. Explain your answer.

_____

_____ [2 marks]

Remember
In questions like this, where you are asked to 'suggest', use the information you are given, as well as your own knowledge, to come up with sensible ideas.

3.

Higher Tier only

The Norfolk Broads are the largest area of protected wetland in Britain. They were originally formed in the Middle Ages when peat was removed for fuel, leaving large pits which later flooded. Many rare bird and insect species now live in the wetlands.

Evaluate the statement 'Wherever possible we should try to undo any environmental damage that humans have caused'. [2 marks]

Worked Example

Sometimes it would help to preserve endangered species if we could reverse the effects of environmental damage, for example by replanting forests where there has been deforestation. However, sometimes humans have been responsible for creating new habitats, like the Broads, and if we tried to reverse this then that could also endanger many species that live there now.

This answer gains 2 marks. The student has evaluated the statement by making points both against and in support of it. Suitable examples have also been used.

Effects of human activities

1.

a What is meant by the term 'biodiversity'?

_____ [1 mark]

b Explain why it is important to maintain a high level of biodiversity.

_____

_____

_____ [2 marks]

2. The human population is causing more pollution than ever before.

Explain why.

_____

_____

_____

_____ [4 marks]

3. Large areas of rainforest have been cut down in tropical areas.

Give **two** reasons why.

1 _____ [1 mark]

2 _____ [1 mark]

> **Hint**
> In questions like this, if you can, try to think of quite different reasons. Otherwise you might give two answers that are alternatives of the same marking point.

4. Many areas of peat bog have been destroyed to remove the peat.

a Give **one** use for peat.

_____ [1 mark]

b Describe and explain the environmental impact of using peat.

_____

_____

_____

_____

_____ [4 marks]

Global warming

1.   Name **two** gases that contribute to global warming.

1 _____ [1 mark]

2 _____ [1 mark]

2.   The graph shows the average global temperature change (compared with the 1880 average global temperature) between 1880 and 2006.

Maths   **a**   Suggest why the temperatures have been shown as temperature differences and **not** as the actual temperatures.

_____

_____ [2 marks]

Maths   **b**   Describe what the graph shows and evaluate to what extent it provides evidence for global warming.

_____

_____

_____

_____

_____

_____

_____

_____ [6 marks]

3. Give **two** possible **biological** consequences of global warming.

> **Remember**
> You need to read the question carefully. The question asks about **biological** consequences, so mentioning consequences such as rising sea levels, or melting ice caps, will not on their own gain any marks.

1 _____ [1 mark]

2 _____ [1 mark]

Maintaining biodiversity

• •

1. Explain how each of the following can help maintain biodiversity.

a Breeding programmes for endangered species:

_____

_____ [2 marks]

b Recycling resources rather than dumping them as waste in landfill sites:

_____

_____ [2 marks]

c Planting trees to turn land back into woodland:

_____

_____ [2 marks]

d Protecting rare habitats, such as marsh land or sand dunes:

_____

_____ [2 marks]

2. After the Second World War many small farms in Britain were joined to form fewer, but larger, farms. Small fields, which used to grow different crops, were joined to grow one type of crop on a large scale. This type of farming is known as monoculture.

a Explain why monoculture has led to a decrease in biodiversity in the countryside. [2 marks]

Worked Example

Fewer different crop plant species means there are fewer insect species feeding on them and fewer birds and other species who feed on the insects. This reduces biodiversity.

This is a good answer and gains the 2 marks. The student could also have mentioned the fact that those insect species that feed on the crop plants often increase in number becoming a pest. This is because many of their competitors or predators have disappeared with the reduction in biodiversity.

b Biodiversity in areas of monoculture can be increased by reintroducing hedgerows and field margins containing wildflowers.

Explain how this helps to increase biodiversity.

_____

_____

_____ [2 marks]

Biomass in an ecosystem

1. Explain what is meant by the term 'biomass'.

Synoptic

_____ [1 mark]

2. This a food chain found in park land.

grass ⟶ rabbits ⟶ foxes

The table shows the biomass at each trophic level.

Trophic level	Total biomass in kg/km²
Foxes	20
Rabbits	160
Grass	2200

Maths **a** Use the information in the table to construct a pyramid of biomass for the food chain.

[2 marks]

Maths **b** Calculate the efficiency of biomass transfer between the trophic levels.

Give your answers to 1 decimal place.

_____

_____

Efficiency of transfer between grass and rabbits = _____ % [1 mark]

Efficiency of transfer between rabbits and foxes = _____ % [1 mark]

> **Maths**
> To calculate the efficiency of transfer you need to use this formula:
>
> $$\text{efficiency} = \frac{\text{biomass after the transfer}}{\text{biomass before the transfer}} \times 100$$
>
> It is a good idea, as a check when doing calculations, to look at the numbers and estimate roughly what sort of answer you should be getting.

c Give **two** reasons why the transfers are **not** 100% efficient.

1 _____ [1 mark]

2 _____ [1 mark]

Synoptic **d** Suggest a reason for the difference in efficiency between the two transfers.

_____

_____

_____

[2 marks]

3. Explain why most food chains only consist of three or four trophic levels and **not** more.

_____

_____ [1 mark]

Food security

1. Explain the term 'food security'.

_____ [1 mark]

2. Put ticks in the boxes to show which of the following would threaten a country's food security.

☐ Being at war.

☐ Drought.

☐ Reduced birth rate.

☐ Vaccination of farm animals against infection. [2 marks]

3. It is important that sustainable methods are used to produce food.

Explain how quotas and control of net size can help make sure that fishing in the oceans is sustainable.

Fishing quotas:

_____

_____

Control of net size:

_____

_____ [4 marks]

4. **a** Describe and explain how the efficiency of food production from animals can be improved using intensive farming methods. [4 marks]

Worked Example

Some animals can be kept inside where they are warm and where they cannot move around much. They can be fed high-protein foods to increase growth.

> This answer gains 3 marks. The student has described several intensive farming techniques but has only explained the use of high-protein foods. To gain full marks they also need to explain that keeping animals warm or restricting their movement means they do not need to release as much energy in respiration as they would otherwise. This means the animals grow bigger at a faster rate.

b Some people object to the use of intensive farming methods to improve food production from animals.

Suggest **two** of their reasons.

1 _____ [1 mark]

2 _____ [1 mark]

Role of biotechnology

1. The fungus *Fusarium* can be grown to produce mycoprotein. This is a protein-rich food which is often textured and flavoured to taste like meat, although it contains less fat than meat. Mycoprotein has been produced for over 30 years and is used in many food products.

Explain why some people would prefer to eat mycoprotein rather than meat.

_____

_____

_____ [2 marks]

2. Golden rice is a type of genetically modified rice. It has been modified to contain a substance that the human body can use to make vitamin A. It was produced to combat vitamin A deficiency, which each year causes blindness or death for hundreds of thousands of people.

Synoptic **a** What does 'genetically modified' mean?

_____ [1 mark]

135

b Suggest why rice was chosen to be the food product that was genetically modified to combat vitamin A deficiency.

_____

_____

_____ [2 marks]

3. Mycoprotein and golden rice are two examples of biotechnology.

a Explain what is meant by the term 'biotechnology'.

_____

_____ [2 marks]

b Suggest why the use of biotechnology to produce food will probably become more common in the future.

_____

_____

_____ [2 marks]

> **Hint**
> Biotechnology is developing all the time so do not be surprised if you get a question about an unfamilar example of biotechnology. Everything you need to answer will be in the question or in what you have studied.

Plant and animal cells (eukaryotic cells)

1. **a** A [1 mark]

   b E/ribosomes. [1 mark]

   c Cytoplasm [1 mark]; where chemical reactions take place. [1 mark]

2. **a** Cellulose. [1 mark]

   b Protection from pathogen attack [1 mark]; structure/shape/support [1 mark]; allows substances through non-selectively. [1 mark]

3. **a** Mitochondrion. [1 mark]

   b Muscle/root hair/nerve/any other valid cell. [1 mark]

   c Worked example – details given in workbook.

Bacterial cells (prokaryotic cells)

1. Bacterial cell walls are **not** made of cellulose [1 mark]; bacteria contain loops of DNA called plasmids. [1 mark]

2. **a** Flagellum [1 mark]; movement. [1 mark]

   b Prokaryota/prokaryote. [1 mark]

   c (Slime) capsule. [1 mark]

   d Protects from environmental hazards. [1 mark]

3. 2.5/two and a half (hours). [1 mark]

4.

Level 2 [3–4 marks] More than **one** similarity and more than **one** difference.
Level 1 [1–2 marks] At least **one** similarity and **one** difference.
No relevant content [0 marks]

Indicative content:

Similarities:

Both have:
- cell wall
- cell membrane
- cytoplasm
- genetic material
- ribosomes.

Differences:
- plant cell wall made of cellulose but bacterial cell wall is not
- plant cell larger in size/can be seen under light microscope OR bacterial cell smaller in size/ cannot be seen under light microscope.

Only bacterial cell has:
- loop of DNA/plasmid
- slime capsule
- flagellum/flagella.

Only plant cell has:
- nucleus (containing genetic material)
- mitochondria
- chloroplasts
- vacuole.

Size of cells and cell parts

1. Ribosome, mitochondrion, nucleus, plant cell vacuole. [all correct = 2 marks, 2 or 3 in correct sequence = 1 mark]

2. 0.12 (mm) [1 mark]

3. Magnification = $20 \times 40 = 800$ [1 mark]; real size = $8/800 = 0.01$ mm [1 mark]; $0.01 \times 1000 = 10$ μm. [1 mark]

4. Conversion of 60 mm to 60 000 μm [1 mark]; magnification = $60\ 000/150 = \times 400$. [1 mark]

5. **a** Iodine/methylene blue/any other valid stain. [1 mark]

   b Any **one** from: to protect specimen from contact with the objectives lens; to keep the specimen pressed flat; to prevent dehydration. [1 mark]

   c To prevent air bubbles. [1 mark]

   d x200 [1 mark]

   e Worked example – details given in workbook.

Electron microscopes

1. Resolving power [1 mark]; lower [1 mark]; organelles. [1 mark]

2. Electron microscope because ribosomes are too small to see through a light microscope/need a high resolving power to see them. [1 mark]

3.

Level 2 [3–4 marks] At least **two** advantages and **two** disadvantages.
Level 1 [1–2 marks] At least **one** advantage and **one** disadvantage.
No relevant content [0 marks]

Indicative content:

Advantages of light microscope:
- portable/easy to use
- specimens can be live/specimens easy to prepare
- no specialist training required
- relatively cheap
- natural colour of specimen can be observed.

Disadvantages of light microscope:
- lower magnification/lower resolving power
- not easy to see details inside cells/sub-cellular structure.

4. **a** Magnification is how many times bigger the image is than the object [1 mark]; resolving power is how sharp the image is/the ability to distinguish very close objects as separate images. [1 mark]

   b It lets us see finer details of sub-cellular structures/ qualified example, e.g. internal structure of mitochondria [1 mark]; not visible under a light microscope. [1 mark]

Growing microorganisms

1. Binary fission. [1 mark]

2. **a** (Antibiotic) A [1 mark]; largest clear zone (of inhibition). [1 mark]

   b Any **two** from: same concentrations/same size of paper disc/soaked in antiseptic for same amount of time/same type of agar/same bacterial strain/incubate for same length of time. [2 marks]

   c To provide nutrients for the bacteria. [1 mark]

3. Correct formula used/area = πr^2 [1 mark]; correct conversion to cm/cm^2 [1 mark]; 12.6 (cm^2). [1 mark]

Cell specialisation and differentiation

1. **a** A: red blood cell [1 mark]; B: nerve cell/neurone [1 mark]; C: (ciliated) epithelial cell [1 mark]; D: root hair cell. [1 mark]

   b Large surface area [1 mark]; to absorb water/mineral ions. [1 mark]

   c Oxygen. [1 mark]

2. **a** To absorb nutrients. [1 mark]

   b A: microvilli [1 mark]; B: mitochondria. [1 mark]

   c A: increased surface area (for absorption) [1 mark]; B: respiration/energy release for active transport of nutrients. [1 mark]

3.

Level 2 [3–4 marks] Definition of differentiation **and** a difference between differentiation in plant and animal cells.
Level 1 [1–2 marks] Definition of differentiation **or** a difference between differentiation in plant and animal cells.
No relevant content [0 marks]

Indicative content:

Cell differentiation:

- as cells divide/organism develops, cells form different types to take on different roles
- cells acquire different sub-cellular structures
- cells become specialised.

Difference between plant and animal cells:

- most types of animal cell differentiate at an early stage of life
- many types of plant cell retain the ability to differentiate throughout life.

Cell division by mitosis

1. Parent; daughter; daughter; parent; mitosis. [all correct = 4 marks; 4 correct = 3 marks; 3 correct = 2 marks; 2 correct = 1 mark]

2. Chromosomes are made of a chemical called DNA [1 mark]; different types of organism have different numbers of chromosomes in their cells. [1 mark]

3. Any **two** from: ribosomes; mitochondria; DNA/chromosomes; other valid answers. [2 marks]

4. **a** Stage 3. [1 mark]

   b $(20/240) \times 100$ [1 mark]; = 8.3 (%) (to 2 s.f.). [1 mark]

   c $(30/240) \times 12 \times 60$ [1 mark]; = 90 minutes. [1 mark]

5.

Level 3 [5–6 marks] All three stages described.
Level 2 [3–4 marks] Two stages described.
Level 1 [1–2 marks] One stage described.
No relevant content [0 marks]

Indicative content:

- (Before cell divides/in first stage) DNA replicates; increase in number of sub-cellular structures, e.g. ribosomes and mitochondria.
- (In second stage) one set of chromosomes is pulled to each end of the cell; nucleus divides.
- (In third/final stage) cytoplasm and cell membranes divide; to form two identical cells.

Stem cells

1. An undifferentiated cell [1 mark]; which can become other types of cells/can become specialised/can differentiate/is pluripotent. [1 mark]

2. Some people object to destroying human embryos. [1 mark]

3. **a** Meristems/root tips/shoot tips. [1 mark]

   b Embryonic stem cells only found in embryos/at embryonic stage but plant stem cells found throughout life of plant. [1 mark]

4. **a** Embryo is potential life/destroying or damaging potential life/the cord would have been discarded anyway. [1 mark]

   b Advantage: perfect tissue match/difficult to find donors/no danger of rejection/no need for immunosuppressant drugs. [1 mark]

   Disadvantage: if health condition is genetic then stem cells have same faulty gene/DNA. [1 mark]

5.

Level 2 [3–4 marks] Considers at least **two** risks and **two** benefits.
Level 1 [1–2 marks] Considers at least **one** risk and **one** benefit.
No relevant content [0 marks]

Indicative content:

Benefits:

- to cure illnesses/valid examples; to reduce suffering/improve quality of life; little risk of rejection if from same person.

Risks:

- unknown long-term side effects; chance of rejection if not from the same person; moral/ethical objections.

Diffusion in and out of cells

1. Worked example – details given in workbook.

2. **a** Arrow inwards; higher concentration outside/lower concentration inside/down the concentration gradient. [1 mark]

   b Any **one** from: greater concentration gradient; increased surface area; shorter diffusion pathway. [1 mark]

3. **a** Cell A. [1 mark]

   b Diffusion occurs fastest with the largest difference between the concentrations inside and outside the cell/greatest concentration gradient. [1 mark]

4.

Level 3 [5–6 marks] There is a description of a process involving diffusion occurring in animals that is correctly linked to a substance **and** a description of a process involving diffusion occurring in plants that is correctly linked to a substance.
Level 2 [3–4 marks] At least one example of a substance is given **and** correctly linked to a process involving diffusion in either animals or plants.
Level 1 [1–2 marks] An example is given of a named substance **or** a process involving diffusion **or** there is an idea of why diffusion is important.
No relevant content [0 marks]

Indicative content:

Importance of diffusion:

- to take in substances for use in cell processes
- to remove products from cell processes.

Examples of processes and substances:

- for gas exchange/respiration: O_2 in/CO_2 out
- for gas exchange/photosynthesis: CO_2 in/O_2 out
- food molecules absorbed: glucose, amino acids, etc.
- water absorption in the large intestine
- water lost from leaves/transpiration
- water absorption by roots.

Description of processes might include:

- movement of particles/molecules/ions
- through a partially permeable membrane
- down a concentration gradient.

Exchange surfaces in animals

1. Any **one** from: thin/short diffusion path; large surface area; efficient blood supply/concentration gradient. [1 mark]

2. Oxygen arrow from air in alveolus to red blood cell [1 mark]; carbon dioxide arrow from plasma (in capillary) to air in alveolus. [1 mark]

3. **a** A = $(3 \times 3 \times 6 =)$ 54 [1 mark]; B = $(3 \times 3 \times 3 =)$ 27 [1 mark]; C = $(5 \times 5 \times 6 =)$ 150 [1 mark]; D = $(150 : 125 =)$ 1.2 : 1. [1 mark]

   b Cube 3/5 × 5 × 5. [1 mark]

   c Smallest surface area to volume ratio/longest distance to centre of cube. [1 mark]

4.

Level 3 [5–6 marks] Explanation of **three** adaptations.
Level 2 [3–4 marks] Explanation of **two** adaptations.
Level 1 [1–2 marks] Explanation of **one** adaptation.
No relevant content [0 marks]

Indicative content:

- Many gill filaments; for a large surface area.
- Gills are thin; so short diffusion pathway.
- Good blood supply; to maintain (large) concentration gradient.
- Water continually flows over them; to maintain (large) concentration gradient.

Osmosis

1. **a** Osmosis. [1 mark]

   b Allows movement through of *some* substances/particles/molecules/ions. [1 mark]

2. **a** Any **two** from: cells in solution B have a smaller vacuole than cells in solution A; cytoplasm of cells in solution B has shrunk (compared with cells in solution A); cells in solution B have gap between cytoplasm/cell membrane and cell wall; cell walls of cells in solution B curve inwards; cells in solution B are flaccid and cells in solution A are turgid. [2 marks]

   b Water will move in [1 mark]; cells will swell/cells will burst. [1 mark]

3. Worked example – details given in workbook.

4. **a** Concentration of sugar in the bag was higher than in the solution (or vice versa)/higher water concentration outside the bag (or vice versa)/ [1 mark]; so water moved in/water moves down its concentration gradient [1 mark]; by osmosis. [1 mark]

   b B [1 mark]

   c Concentration of solution is close to the concentration in the bag/to 5.0% [1 mark]; rate of osmosis/diffusion is low/relatively little water needs to move to equalise the concentrations inside and outside. [1 mark]

   d Allows comparisons [1 mark]; (between) different starting masses/weights. [1 mark]

Active transport

1. **a** Magnesium/nitrate/any other valid answer. [1 mark]

   b (Active transport) requires energy [1 mark]; (movement) against concentration gradient. [1 mark]

2. **a** Diffusion. [1 mark]

   b Active transport. [1 mark]

   c Mineral ions needed for healthy (plant) growth/allow specific example, e.g. nitrates needed to make proteins or magnesium needed to make chlorophyll. [1 mark]

   d 'Hair' or projection [1 mark]; increases surface area (to volume ratio). [1 mark]

3. **a** Gut/small intestine. [1 mark]

   b Glucose/sugar/(accept amino acids). [1 mark]

   c Higher concentration (of sugar) in blood/lower concentration in gut [1 mark]; will prevent diffusion into blood. [1 mark]

Section 2: Organisation

Digestive system

1. **a** Oesophagus. [1 mark]

   b Pancreas. [1 mark]

2. **a** Hydrochloric acid/protease (enzymes). [1 mark]

   b Hydrochloric acid – to kill harmful microbes/provide optimum pH for stomach enzymes/digestion **or** protease (enzymes) – digest protein. [1 mark]

3. A group of tissues [1 mark]; performing a specific function. [1 mark]

4. Peristalsis/contractions of the muscular wall. [1 mark]

5. Worked example – details given in workbook.

6.
Level 2 [3–4 marks] Detailed description of the link between damaged villi **and** being underweight.
Level 1 [1–2 marks] Description of the effect of damaged villi.
No relevant content [0 marks]
Indicative content:
• Flattened/no villi which reduces surface area. • So, there is reduced absorption of digested food. • Reduction in food absorption leads to being underweight. • Lack of proteins/amino acids causes reduced growth. • Lack of fat/lipid/glycerol and fatty acids causes reduced weight.

Digestive enzymes

1. Catalyst [1 mark]; stomach [1 mark]; fat. [1 mark]

2. Any **four** from: amylase; produced in mouth/salivary glands; and pancreas/small intestine; breaks down starch (to simple sugars); to maltose. [4 marks]

3. Add Benedict's solution [1 mark]; heat in a water bath [1 mark]; positive result is a change (from a blue colour) to a brick-red colour. [1 mark]

Factors affecting enzymes

1. **a** This shows the time taken for all the starch to be digested. [1 mark]

   b Temperature affects enzyme activity/rate of enzyme-controlled reactions **or** to make sure changing temperature did not affect rate of reaction. [1 mark]

   c 77 [1 mark]; 140. [1 mark]

   d Any **three** from: (initially reaction is) faster as pH increases; fastest at pH7; slower as pH increases from pH7; pH7 is optimum. [3 marks]

   e Any **three** from: extremes of pH change structure/shape of (enzyme) active site; substrate/starch can no longer fit in active site; amylase/enzyme can no longer function as a catalyst/break down starch; irreversible change. [3 marks]

   f Use smaller intervals of pH, e.g. pH7.5 [1 mark]; to more accurately identify the optimum pH [1 mark] **or** use shorter time intervals, e.g. 5 seconds [1 mark]; to more accurately measure the time taken for the reaction [1 mark] **or** further repeats [1 mark]; so you can calculate more accurate means. [1 mark]

The heart and blood vessels

1. **a** A = aorta [1 mark]; B = left ventricle. [1 mark]

   b Arrow(s) on right side of heart (left side of diagram) showing blood moving from vena cava into right atrium, right ventricle, pulmonary artery. [1 mark]

2. **a** In the right atrium. [1 mark]

   b Regulate the heart rate/trigger contraction of heart. [1 mark]

3. A system where blood flows from heart to lungs and back [1 mark]; (then) blood flows from heart to body and back. [1 mark]

4. Wall of capillary is (very) thin; one cell thick; allows exchange of materials with body tissues **or** large number of small capillaries; provides a large surface area; for exchange of materials. [3 marks]

5. 1600 ml = 1.6 l; rate = 1.6/4 = 0.4 (l/min) [2 marks]

6. Worked example – details given in workbook.

Blood

1. Any **two** from: water; carbon dioxide; glucose; amino acids; proteins; fatty acids; glycerol; vitamins; ions/named ions; urea; antibodies; antitoxins; cholesterol; hormones/named hormones. [2 marks]

2. **a** 55(%) [1 mark]

   b Red blood cells. [1 mark]

3. White blood cells fight infection/pathogens [1 mark]; so, a low white cell count means a greater risk of infection. [1 mark]

4. Any **three** descriptions [3 marks] and corresponding explanations. [3 marks]

Description	Explanation
small/flexible	so they can pass through capillaries
biconcave disc shape	for large surface area/efficient diffusion

| contain haemoglobin | haemoglobin binds to oxygen |
| no nucleus | to increase space for haemoglobin |

Heart and lungs

1. **a** A = trachea/wind-pipe [1 mark]; B = lung [1 mark]; C = (right) bronchus. [1 mark]

   b (C-shaped) cartilage rings [1 mark]; keep the trachea/wind-pipe open. [1 mark]

2. Large surface/area [1 mark]; thin/one cell thick/short distance (from air to blood) [1 mark]; many capillaries/ blood vessels/good blood supply. [1 mark]

3. Worked example – details given in workbook.

Coronary heart disease

1. **a** Females 85+. [1 mark]

   b 5750 : 1518 = 3.78 : 1 [1 mark] **but** 3.8 : 1 [2 marks]

2. Fatty deposits/material in (coronary) arteries [1 mark]; reduces flow/narrows/blocks [1 mark]; decreases oxygen supply to heart muscle. [1 mark]

3. Statins – lower cholesterol [1 mark]; decrease build-up of fat (in arteries). [1 mark]

   Stents – open up/widen arteries [1 mark]; to open blocked parts/increase flow of blood. [1 mark]

4.

| Level 2 [3–4 marks] An evaluation based on at least two advantages **and** two disadvantages. |
| Level 1 [1–2 marks] An evaluation based on at least two advantages **or** two disadvantages **or** one advantage and one disadvantage. |
| No relevant content [0 marks] |

Indicative content:

Advantages:

- improved circulation provides more oxygen
- more energy released
- more active life/not so tired
- less risk of heart attack.

Disadvantages:

- risk of infection
- operation/recovery takes long time
- clots may form and block blood vessels
- clots may cause heart attacks/strokes.

Risk factors for non-infectious diseases

1. 5.8×10^{-2} [1 mark]

2. Do more exercise [1 mark]; lose weight/eat less fat/eat less salt/eat healthier diet [1 mark]; stop smoking/do not smoke. [1 mark]

3. A risk factor is something that increases the risk of a disease [1 mark]; a causal mechanism is an explanation for why a particular factor increases the risk of a disease. [1 mark]

4. **a** As BMI increases (in males/females/both), so does the probability of developing Type 2 diabetes/and vice versa. [1 mark]

   b A higher BMI is due to an increase in mass/weight/ obesity [1 mark]; being obese/overweight is a risk factor for Type 2 diabetes/causes Type 2 diabetes. [1 mark]

Cancer

1. Exposure to UV light **and** old age **and** smoking [2 marks] **or** two correct. [1 mark]

2. **a** Any **three** from: detection rate increased (until 2015) and decreased in 2016; decrease in detection rate was more rapid than increase; death rate getting lower each year; greatest rate of decrease in death rate was between 2015 and 2016. [3 marks]

   b Detection rate might have increased as more people/ men responded to health campaigns by seeking medical help if they suspected cancer **or** detection rate might have decreased as fewer people/men responded to health campaigns by seeking medical help if they suspected cancer. [1 mark]; death rate has decreased because of better/advanced methods of diagnosis/early detection/better treatments. [1 mark]

3. Worked example – details given in workbook.

Leaves as plant organs

1. **a** **A** – vascular bundle/vein/xylem **and** phloem [1 mark]; **B** – stoma (allow stomata). [1 mark]

   b Part **A** – supports the leaf **or** transports water to leaf/sucrose away from leaf [1 mark]; part **B** – controls gas exchange/controls water loss/ transpiration. [1 mark]

2. **a** Air spaces [1 mark]; allow gas exchange/diffusion of oxygen/diffusion of carbon dioxide/diffusion of water vapour. [1 mark]

   b No chloroplasts/transparent [1 mark]; (protects leaf surface but) allows light to enter leaf/palisade layer. [1 mark]

3.

| Level 2 [3–4 marks] At least **two** adaptations described and explained. |
| Level 1 [1–2 marks] At least **one** adaptation described and explained **or** two adaptations described. |
| No relevant content [0 marks] |

Indicative content:

- Position near upper surface – to absorb maximum sunlight.
- Contains many chloroplasts – to absorb light energy.
- Long thin shaped cells – so each cell can both receive sunlight from above and carbon dioxide from below.

Transpiration

1. **a** Guard cells. [1 mark]

   b Stomata are open. [1 mark]

   c Worked example – details given in workbook.

2.
Level 3 [5–6 marks] Detailed description and explanation of the process of transpiration.
Level 2 [3–4 marks] Detailed description of the process of transpiration.
Level 1 [1–2 marks] Outline description of the process of transpiration.
No relevant content [0 marks]

   Indicative content:
   - Loss of water from plant leaves.
   - Water moves up a plant to the leaves.
   - Water moves through xylem.
   - In one direction only.
   - Xylem made of hollow vessels with lignified walls.
   - Water evaporates from cells inside leaf and diffuses out of the leaf.
   - Water (vapour) diffuses out through stomata.
   - More water drawn up xylem from roots to replace it.
   - Transpiration stream carries water and mineral ions up the plant.

Translocation

1. **a** Movement of food/sugar through a plant [1 mark]; from leaves/storage regions to rest of plant. [1 mark]

   b Tubes of elongated cells [1 mark]; pores in end walls allow cell sap to move from one cell to another. [1 mark]

2. Worked example – details given in workbook.

3.
Level 3 [5–6 marks] Detailed description of the movement of mineral ions and sugar.
Level 2 [3–4 marks] Detailed description of the movement of mineral ions or sugar.
Level 1 [1–2 marks] Brief description of the movement of mineral ions or sugar.
No relevant content [0 marks]

   Indicative content:

   Mineral ions:
   - taken up from soil by root hair cells
   - by active transport
   - (active transport) uses energy from respiration
   - carried up plant in xylem tissue
   - in one direction only
   - carried in transpiration stream
   - dissolved in water.

Sugars:
- sugar/glucose made in leaves during photosynthesis
- sugars moved from leaves to other parts of plant
- to be stored
- sometimes stored as starch
- (or) used for growth/energy
- some sugars moved from storage regions to growing regions
- moved through phloem tissue
- in any direction/up and down plant
- moved by translocation
- dissolved in water.

Section 3: Infection and response

Microorganisms and disease

1. Diseases caused by pathogens/can be spread from one person to another. [1 mark]

2. Malaria; measles. [2 marks]

3. Pathogens are microorganisms that cause infectious disease. [1 mark]

4. Any **three** from: spread by direct contact; by water; by air; by a vector; by contaminated food. [3 marks]

5. Worked example – details given in workbook.

Viral diseases

1. **a** Mosaic pattern of discolouration/patchy chlorophyll. [1 mark]

   b Worked example – details given in workbook.

2.
Level 2 [3–4 marks] At least **two** similarities **and two** differences.
Level 1 [1–2 marks] At least **two** similarities **or two** differences **or one** similarity and **one** difference.
No relevant content [0 marks]

   Indicative content:

   Similarities:
   - both viral
   - neither respond to antibiotics
   - initial cold-/flu-like symptoms
   - both infectious (from first symptoms)
   - both can be fatal.

   Differences:
   Measles:
   - spread by droplet infection/coughs/sneezes
   - cold-like symptoms/fever/rash
   - complications such as blindness/meningitis.

HIV:
- spread through body fluids/unprotected sexual contact/sharing needles
- remains hidden/becomes active after months or years/leading to AIDS
- affects immune system/white blood cells
- antiretroviral (drugs) can slow development of AIDS.

3. a 2011–2012 [1 mark]

   b The graph shows only those cases confirmed in a laboratory/not every case is confirmed in a laboratory. [1 mark]

   c Unvaccinated people may have caught measles when travelling abroad (to countries where measles still occurs) **or** caught it from an infected person who has come from abroad. [1 mark]

Bacterial diseases

1. Bacteria have cell walls. [1 mark]

2. a Any **three** from: both show an overall increase; males higher than females; males rising faster than females; both show rising and then falling, and then rising again. [3 marks]

   b $((33.1 - 22.7)/22.5) \times 100 = 46.2$ [1 mark]; **but** 46%. [2 marks]

   c Any **two** from: more people getting tested for gonorrhoea; new diagnostic tests; antibiotic resistance; unsafe sex/not using condoms. [2 marks]

3.

Level 2 [3–4 marks] At least **two** similarities **and two** differences.
Level 1 [1–2 marks] At least **two** similarities **or two** differences **or one** similarity and **one** difference.
No relevant content [0 marks]
Indicative content:

Similarities:
- both caused by bacteria
- both can be treated by antibiotics.

Differences:

Gonorrhoea:
- sexually transmitted
- symptoms include discharge/pain on urinating
- prevented by using condoms.

Salmonella:
- ingested in food
- symptoms include fever/abdominal cramps/vomiting/diarrhoea
- controlled by vaccinating poultry.

Malaria

1. Protist/*Plasmodium*. [1 mark]

2. Vectors. [1 mark]

3. Any **one** from: (recurrent) fever; headaches; muscle pains; vomiting; diarrhoea. [1 mark]

4.

Level 2 [3–4 marks] Description and explanation of at least **two** ways malaria can be controlled.
Level 1 [1–2 marks] Description and explanation of at least **one** way malaria can be controlled.
No relevant content [0 marks]
Indicative content:

Prevention of breeding:
- kill larvae by using chemicals/insecticides/poisons
- introduce sterile males
- destroy breeding grounds, e.g. by draining areas of stagnant water
- empty/drain/cover all things that hold water
- explanation: these all prevent mosquitoes from breeding.

Prevention of biting:
- use chemicals/sprays/repellents
- use (mosquito) nets
- explanation: these all prevent mosquitoes from biting.

5. (1×10^6) divided by $(5 \times 10^8) = 0.002$ [1 mark]; **but** 0.2 (%). [2 marks]

Human defence systems

1. a Provides a barrier; if broken, a scab is formed. [2 marks]

   b Mucus traps pathogens; cilia move mucus/pathogens up/out of airways. [2 marks]

   c Produces hydrochloric acid; kills pathogens ingested/in food or drink. [2 marks]

2. More likely to suffer from other infectious diseases; white blood cells not (as) able to defend against pathogens. [2 marks]

3.

Level 3 [5–6 marks] Detailed description of at least **three** ways white blood cells defend against pathogens.
Level 2 [3–4 marks] Detailed description of at least **two** ways white blood cells defend against pathogens.
Level 1 [1–2 marks] Detailed description of at least **one** way white blood cells defend against pathogens **or** a partial description of several ways white blood cells defend against pathogens.
No relevant content [0 marks]
Indicative content:

Method 1:
- phagocytosis/ingest pathogens
- pathogens digested/broken down by enzymes.

Method 2:
- antibody production
- antibodies bind to pathogens/antigens leading to their destruction
- correct reference to memory cells.

Method 3:
- antitoxin production
- antitoxins neutralise toxins produced by pathogens.

Vaccination

1. Inactive/weakened/attenuated/dead pathogen. [1 mark]

2. Any **four** from: (dead/inactive pathogens) stimulate white blood cells to produce antibodies; memory cells formed; if same pathogen re-enters body, white blood cells/memory cells respond/reproduce quickly; to produce the correct antibodies again; preventing infection. [4 marks]

3. Vaccine does not affect pathogens **or** white blood cells will already be responding to the pathogens. [1 mark]

4. Pathogen/virus mutates/changes; white blood cells/ immune system do not recognise new strains; need new vaccines for each new strain. [3 marks]

5. Worked example – details given in workbook.

Antibiotics and painkillers

1. Antibiotics – gonorrhoea.
   Painkillers – headache; toothache. [3 marks]

2. Painkillers will ease pain but they won't kill the pathogens causing the pain. [1 mark]

3. Viruses live inside cells; (so) it is difficult to develop drugs that kill viruses without also damaging the body's cells/tissues. [2 marks]

4. **a** Not all infections are caused by bacteria; antibiotics kill bacteria but not viruses/antibiotics won't treat diseases caused by viruses; overuse of antibiotics has led to emergence of resistant bacterial strains/to avoid overuse of antibiotics and emergence of resistant bacterial strains; many infections will clear up without treatment/because of body's own defences. [4 marks]

   b If infection is (definitely) caused by bacteria; (and) if infection is severe/may not clear up on its own/may lead to further complications. [2 marks]

Making and testing new drugs

1. Foxgloves; Alexander Fleming; willow. [3 marks]

2. **a** No, some volunteers should be given no drugs/ placebo; for comparison/to see if **A** and **B** really work. [2 marks]

   b Any **two** from: age; gender; ethnicity; severity of pain/how long they had pain before trial; type of pain/illness/site of pain; weight/height; other medical issues. [2 marks]

   c Any **one** from: large number makes results repeatable; can identify anomalous results; pain is subjective so need a large amount of data to have confidence in results; if they had used small number the results might not have been representative. [1 mark]

   d To find whether the drug is toxic/safe to use on humans. [1 mark]

3. **a** Everything that's in the actual drug except the active ingredient. [1 mark]

   b As a comparison/control for the actual drug. [1 mark]

   c Neither the doctor/scientist carrying out the trial nor the patient/volunteer knows if they are using the actual drug or the placebo. [1 mark]

   d To avoid bias (by doctor/scientist/patient/volunteer). [1 mark]

Monoclonal antibodies

1. **a** Correct label to one of the antibodies (Y-shaped structures). [1 mark]

   b Antibodies of (just) one type/all same type. [1 mark]

2. **a** To stimulate production of lymphocytes (which produce antibodies). [1 mark]

   b So hybridoma cells will divide quickly (like tumour cells); (and) produce antibodies (like lymphocytes). [2 marks]

   c Hybridoma cells are clones (identical copies). [1 mark]

3. **a** Monoclonal antibodies are able to target specific antigens on cancer cells; they can carry radioactive substances/toxic drugs/chemicals; they are delivered to cancer cells (only); without harming other cells. [4 marks]

   b Any **two** from: pregnancy tests; measuring levels of hormones/other chemicals in blood; detecting pathogens; identifying specific molecules in a cell/ tissue. [2 marks]

   c They create more side effects than expected/may provoke immune response in patients. [1 mark]

Plant diseases

1. Disease may be communicable/pathogen could be spread by infected tools. [1 mark]

2. **a** A fungus. [1 mark]

   b Loss of green colour/chlorophyll; (means) leaves cannot photosynthesise/make food. [2 marks]

   c By wind; by rain. [2 marks]

   d Worked example – details given in workbook.

3. Antibodies obtained from plasma of animal/rabbit; after it has been injected with the plant virus/isolated plant pathogen antigen. [2 marks]

Identification of plant diseases

1. Any **two** from: refer to gardening manual/website; use testing kits/using monoclonal antibodies; laboratory identification. [2 marks]

2. Any **four** from: (stunted) growth; spots on leaves; areas of decay/rot; growths; (malformed) appearance of leaves/stems/other plant parts; discolouration; presence of pests. [4 marks]

3. **a** Deficiency of magnesium; yellow leaves caused by lack of chlorophyll/magnesium needed to make chlorophyll. [2 marks]

   b Deficiency of nitrate; stunted growth caused by lack of protein/nitrate needed to make protein. [2 marks]

4. Any **two** from: suck sap/remove sugars/nutrients; reduce sugar/energy available for growth; aphids carry viruses/spread disease. [2 marks]

Plant defence responses

1. Hawthorn trees – mechanical; nettle leaves – chemical; basil – physical.

   One/two correct = [1 mark] **but** all correct = [2 marks]

2. Chemical defence; kill/prevent microorganisms from attacking/infecting; smell/taste (of oil) deters insects/herbivores. [3 marks]

3.
Level 2 [3–4 marks] Detailed explanation of **all** the adaptations.
Level 1 [1–2 marks] Brief explanation of most adaptations **or** detailed explanation of some.
No relevant content [0 marks]
Indicative content: • Defences lessen/prevent damage to tree by animal/herbivores. • Spines deter animals eating leaves. • Tannins/chemicals deter animals from eating leaves. • Trees grow tall to prevent leaves being eaten by short animals/named example of animal. • Stinging ants in spines deter animals from eating leaves.

4. Worked example – details given in workbook.

Section 4: Photosynthesis and respiration reactions

Photosynthesis reaction

1. **a** Chlorophyll. [1 mark]
   b Endothermic. [1 mark]
   c Stomata. [1 mark]

2. carbon dioxide + water $\xrightarrow{\text{light}}$ glucose + oxygen
   At least two substances correct = [1 mark] **but** all correct = [2 marks]

3. $6CO_2 + 6H_2O \xrightarrow{\text{light}} C_6H_{12}O_6 + 6O_2$
   At least two formulae correct = [1 mark] **but** all formulae correct = [2 marks] **but** all correct and balanced = [3 marks]

4. **a** Chloroplasts. [1 mark]

   b Palisade (mesophyll) cells; spongy mesophyll cells; guard cells; circle around palisade cells. [4 marks]

5. $(10 \times 10^{10}$ divided by $7 \times 10^{12}) \times 100 = 1.429$ [1 mark] **but** = 1.4(%) [2 marks]

Rate of photosynthesis

1. Increasing the temperature; turning the lights on at night. [2 marks]

2. The amount of photosynthesis occurring over (a specific amount of) time/how quickly photosynthesis occurs. [1 mark]

3. **a** Any **two** from: temperature; carbon dioxide concentration (of water); size of the pondweed. [2 marks]

   b Number of bubbles/amount of oxygen (gas). [1 mark]

   c Any **two** from: as distance between the lamp and the pondweed increases, the number of bubbles per minute decreases/vice versa; as light intensity decreases, the rate of oxygen production decreases/vice versa; as distance increases/light intensity decreases, the rate of photosynthesis decreases/vice versa. [2 marks]

   d Yes, because the number of bubbles is (approximately) proportional to the inverse square of the distance (1/distance^2) **or** the number of bubbles multiplied by the square of the distance = a constant (approximately); e.g. $106 \times 10^2 = 10\,600$ and $48 \times 15^2 = 10\,800$, almost the same. [2 marks]

4. **a** 33 [1 mark]

   b Green; this produces fewest bubbles; so less photosynthesis. [3 marks]

Limiting factors

1. **a** C [1 mark]

   b Carbon dioxide concentration; because increasing the concentration increases the rate of photosynthesis. [2 marks]

   c Light (intensity)/temperature/**not** carbon dioxide; because increasing the carbon dioxide concentration does **not** increase the rate of photosynthesis. [2 marks]

2. **a** Increasing light intensity increases the rate of photosynthesis at first, until it reaches a maximum; increasing temperature increases the rate of photosynthesis. [2 marks]

b At point **X** light intensity is the limiting factor (so different temperatures do not make a difference); at point **Y** temperature is the limiting factor (so different temperatures do make a difference). [2 marks]

c Worked example – details given in workbook.

Uses of glucose from photosynthesis

1. Light; chlorophyll; decrease. [3 marks]

2. **a** Worked example – details given in workbook.

   b Any **three** from: for respiration/energy; to make cellulose; to make starch; to make fat/lipid/oil; to make protein; to build big molecules from small molecules/metabolism. [3 marks]

3. **a** To make protein; for (healthy) growth. [2 marks]

   b Stunted growth; yellow/pale green leaves. [2 marks]

Cell respiration

1. Any **two** from: to build larger molecules/metabolism; movement; keeping warm. [2 marks]

2. (Aerobic respiration) uses oxygen/releases more energy. [1 mark]

3. glucose + oxygen → carbon dioxide + water
   At least two substances correct = [1 mark] **but** all correct = [2 marks]

4. $C_6H_{12}O_6 + 6O_2 \rightarrow 6CO_2 + 6H_2O$
   At least two formulae correct [1 mark] **but** all formulae correct [2 marks] **but** all correct and balanced. [3 marks]

5. **a** 10.8 (mm^3 per minute) [1 mark]

   b Chemical reactions slower/less enzyme activity/ locusts less active. [1 mark]

   c Oxygen (concentration); amount of glucose/ respirable material available. [2 marks]

   d Worked example – details given in workbook.

6.

Level 2 [3–4 marks] Definition of metabolism **and two** examples of metabolic reactions.
Level 1 [1–2 marks] Definition of metabolism **or** examples of metabolic reactions.
No relevant content [0 marks]

Indicative content:

Definition:

- metabolism is the sum of all reactions in cells/body
- uses energy from respiration
- controlled by enzymes.

Examples of metabolic reactions:

- conversion of glucose to starch/cellulose/ glycogen

- formation of lipid molecules from glycerol and (three) fatty acids
- formation of amino acids from glucose and nitrate ions
- synthesis of proteins from amino acids
- breakdown of excess proteins to form urea
- respiration.

Anaerobic respiration

1. No oxygen (used); lactic acid produced instead of carbon dioxide and water; less energy released. [3 marks]

2. **a** Carbon dioxide. [1 mark]

   b Measure volume of gas produced in a unit of time/ per minute/per second. [1 mark]

3. **a** glucose → ethanol + carbon dioxide [1 mark]

   b Wine making/brewing; bread making/baking. [2 marks]

4. Worked example – details given in workbook.

Response to exercise

1. **a** 1105 (kJ/h) [1 mark]

   b 91.7 (%) [1 mark]

   c Requires more energy to move a larger mass/weight **or** respiration is taking place in more cells. [1 mark]

2. Any **three** from: increased heart rate; increased breathing rate; increased breath volume. [3 marks]

3.

Level 3 [5–6 marks] Definition of oxygen debt **and** detailed description of response.
Level 2 [3–4 marks] Definition of oxygen debt **and** partial description of response.
Level 1 [1–2 marks] Partial definition of oxygen debt **or** partial description of response.
No relevant content [0 marks]

Indicative content:

Definition:

- insufficient oxygen to muscles causes anaerobic respiration
- build up of lactic acid
- oxygen debt is the amount of extra oxygen the body needs after exercise to convert lactic acid to carbon dioxide and water.

Response:

- muscles become fatigued/ache because of low tolerance of lactic acid
- lactic acid transported to liver
- lactic acid converted to glucose
- breathing rate stays elevated/does not return to normal until lactic acid broken down.

Homeostasis

1. Body temperature; water content of body. [2 marks]
2. Receptor – detects changes in the environment – eyes and skin; coordination centre – processes information – brain and spinal cord; effector – brings about a response – muscles and glands. [3 marks]
3. Maintains a steady internal environment; maintains optimum conditions needed for body processes/ enzyme reactions. [2 marks]
4. Nervous system; endocrine/hormone system. [2 marks]

The nervous system and reflexes

1. a Sensory (neurone). [1 mark]
   b Synapse. [1 mark]
   c Chemical transmitter/neurotransmitter. [1 mark]
   d Reflex does not involve the brain/go to the brain **or** coordinated <u>only</u> by the spinal cord. [1 mark]
   e Contracts/gets shorter. [1 mark]
   f They are automatic/rapid responses; that protect the body from danger/damage. [2 marks]
2. a 19.5 = [1 mark] **but** 20 = [2 marks]
   b Caffeine decreases reaction time. [1 mark]
   c Just one person/(very) small sample/drink contained other things as well as caffeine. [1 mark]
   d Any **two** from: more repetitions; measure to the nearest mm; use more people; use different amounts of caffeine; compare with a control/ similar soft drink but without caffeine; use different/more time intervals; more accurate measure of reaction time, e.g. computer-generated. [2 marks]

3.

Level 2 [3–4 marks] At least **two** differences described in detail.
Level 1 [1–2 marks] At least **one** difference described in detail **or** more than **one** difference partially described.
No relevant content [0 marks]
Indicative content:

Receptors:
- cells/nerve endings
- in receptor organs/named examples, e.g. skin, eye
- detect changes in the internal or external environment
- convert stimulus to (nerve) impulse
- to central nervous system.

Effectors:
- muscles or glands
- named examples, e.g. salivary gland, radial muscle
- carry out response to stimulus
- convert nerve impulse to action/response
- from central nervous system.

The brain

1. a X = cerebral cortex/cerebrum; Y = cerebellum; Z = medulla (oblongata). [3 marks]
   b Controls conscious thought/personality/language/ memory/intelligence. [1 mark]
   c Controls/coordinates balance/movement/muscle activity. [1 mark]
   d Controls unconscious/automatic activities/heart rate/breathing rate. [1 mark]
2. a Cerebellum. [1 mark]
   b MRI/CT/PET scan. [1 mark]
   c Worked example – details given in workbook.

The eye

1. a **A** = iris; **B** = retina; **C** = optic nerve; **D** = lens. [4 marks]
   b Line to sclera. [1 mark]
2. Carries nerve impulses to brain – optic nerve; protects eye – sclera; refracts light – cornea; senses light – retina.
   One correct = [1 mark] **but** two/three correct = [2 marks] **but** all correct = [3 marks]
3. Pupil gets larger; circular muscles (in iris) relax; radial muscles (in iris) contract. [3 marks]
4. Ciliary muscles relax; suspensory ligaments are pulled tight; lens is pulled thin; so lens only slightly refracts light. [4 marks]

Seeing in focus

1. Become fatter. [1 mark]
2. a Long sight(edness)/hyperopia. [1 mark]
   b Lens in the eye becomes inflexible/cannot change shape (enough)/cannot become fatter; light/image is focused behind/not on retina. [2 marks]
   c Convex/converging (lens). [1 mark]
3. Diagram showing: a concave/diverging lens; light/rays being diverged before entering eye/pupil; light rays meeting/focusing on retina. [3 marks]
4. Worked example – details given in workbook.

Control of body temperature

1. a The person started exercising **or** the surrounding temperature increased; because sweating increased **or** there was a need to sweat to avoid over-heating. [2 marks]
   b $100 \times 2280/2355 = 96.8(\%) = $ [1 mark] **but** 97(%) = [2 marks]
   c Evaporates; transferring heat from body. [2 marks]

2.

Level 3 [5–6 marks] Clear description of the role of the thermoregulatory centre **and** an explanation of at least **two** control mechanisms.
Level 2 [3–4 marks] Clear description of the role of the thermoregulatory centre **and** an explanation of at least **one** control mechanism.
Level 1 [1–2 marks] Partial description of the role of the thermoregulatory centre **or** a partial explanation of at least **one** control mechanism.
No relevant content [0 marks]

Indicative content:

- Temperature receptors in thermoregulatory centre.
- Thermoregulatory centre monitors blood temperature/core body temperature.
- Temperature receptors in skin send impulses to thermoregulatory centre (giving information about skin temperature).
- If blood/core temperature is too low, blood vessels supplying the skin constrict/ vasoconstriction.
- To reduce flow of blood to skin.
- Less heat is transferred/lost from skin.
- Sweat glands stop releasing sweat.
- So, no cooling by evaporation.
- Muscles shiver/contract.
- To release (more) heat energy from respiration.

Hormones and the endocrine system

1. **A** = pituitary; **B** = thyroid; **C** = adrenal gland(s); **D** = pancreas. [4 marks]

2. **a** Pituitary/A. [1 mark]

b Stimulates other glands (to release hormones). [1 mark]

3.

Level 3 [5–6 marks] Clearly describes at least **three** differences.
Level 2 [3–4 marks] Clearly describes at least **two** differences.
Level 1 [1–2 marks] Clearly describes at least **one** difference **or** gives a partial description of several differences.
No relevant content [0 marks]

Indicative content:

Nervous control:

- fast(er)
- short(er)-lived response
- electrical impulses
- carried through nerves/neurones
- chemical conduction through synapses
- involves receptors/nerve endings in sense organs
- action brought about by effectors/muscles/glands
- localised response.

Hormonal control:

- slow(er)
- long(er)-lived response
- hormones are chemicals
- carried through blood/plasma
- to target organs/cells
- named example **and** action
- (can be) widespread response.

Controlling blood glucose

1. **a** Description: pancreas does not produce enough insulin; treatment: insulin injections. [2 marks]

b Description: body cells do not respond to insulin; treatment: controlled diet and exercise. [2 marks]

2. **a** $(3.7 \times 10^6) \times 0.9 = 3.33 \times 10^6/3\,330\,000/3.3$ million. [1 mark]

b Any **one** from: better tests/diagnosis; an increase in incidence of obesity. [1 mark]

3. Worked example – details given in workbook.

Maintaining water and nitrogen balance in the body

1. Exhalation/breathing; sweating. [2 marks]

2. **a** Amino acids. [1 mark]

b Urea. [1 mark]

3. Worked example – details given in workbook.

4.

Level 3 [5–6 marks] Clear and detailed answer explaining how ADH controls the amount of water excreted by the kidneys.
Level 2 [3–4 marks] Clear answer explaining some details of how ADH controls the amount of water excreted by the kidneys.
Level 1 [1–2 marks] Partial answer describing some aspects of the control of the amount of water excreted by the kidneys.
No relevant content [0 marks]

Indicative content:

If blood concentration too high:

- ADH released from pituitary
- travels in blood to kidney (tubules)
- kidney tubules become more permeable to water
- more water is reabsorbed back into blood
- smaller volume/more concentrated urine
- blood concentration falls/returns to normal.

If blood concentration too low:

- ADH not released from pituitary
- kidney tubules become less permeable to water
- less water is reabsorbed back into blood
- larger volume/less concentrated urine
- blood concentration rises/returns to normal.

Hormones in human reproduction

1. (C) E B D A At least three in correct order = [1 mark] **but** all correct = [2 marks]

2. **a**

Level 2 [3–4 marks] Detailed comparison describing at least **one** similarity **and one** difference.
Level 1 [1–2 marks] Comparison describing at least **one** similarity **or one** difference.
No relevant content [0 marks]

   Indicative content:

   Similarities:
   - both hormones show an overall increase
   - neither decreases (until after 38 weeks).

   Differences:

   Progesterone:
   - starts being produced at 4 weeks
   - increases (steadily) from 4 to 38 weeks
   - is higher than oestrogen from 6 weeks onwards.

   Oestrogen:
   - constant/low level from 0 to 20 weeks
   - increases (steadily) from 20 to 38 weeks
   - is higher than progesterone from 0 to 6 weeks
   - rises more steeply than progesterone.

   b Oxytocin; level increases just before birth/peaks at birth. [2 marks]

3. Oestrogen stimulates production of LH; oestrogen inhibits production of FSH; FSH stimulates production of oestrogen. [3 marks]

Contraception

1. **a** Oestrogen; progesterone. [2 marks]

   b Any **two** from: less chance of forgetting (pills)/lasts for 3 years; reduces heavy/painful periods; less chance of headaches; less chance of (breast) cancer; less risk of blood clots. [2 marks]

   c Any **two** from: painful to insert/uncomfortable/risk of infection; woman can't take it out/needs to be removed if woman wants to become pregnant; can cause irregular periods. [2 marks]

2.

Level 2 [3–4 marks] At least **two** points of each method discussed.
Level 1 [1–2 marks] At least **one** point of each method discussed.
No relevant content [0 marks]

   Indicative content:

   Diaphragm/cap/female condom:
   - hormone free
   - more complicated to fit
   - slightly less effective in preventing pregnancy
   - stops sperm entering uterus
   - no STI protection for diaphragm/cap
   - STI protection for female condom
   - diaphragm/cap stays in after intercourse/can be reused
   - new female condom used each time.

   Male condom:
   - hormone free
   - easy to fit
   - slightly more effective in preventing pregnancy
   - stops sperm entering uterus
   - STI protection
   - new male condom used each time.

Using hormones to treat human infertility

1. **a** To stimulate several eggs to mature. [1 mark]

   b In a dish/laboratory. [1 mark]

   c Not all embryos will successfully develop/low success rate. [1 mark]

   d All/many may develop successfully/multiple pregnancy; health risk to the babies/babies are likely to be small/premature; health risk to mother. [3 marks]

   e Any **two** from: may cause (physical/emotional) stress (during treatment); cost of treatment; issue of what to do with unused embryos/religious objections. [2 marks]

   f Woman does not have ovaries/uterus **or** any other valid answer. [1 mark]

Negative feedback

1. Detecting a change and responding to reverse the change. [1 mark]

2. Worked example – details given in workbook.

3. **a** Adrenal gland(s). [1 mark]

   b Produced in times of stress/fear/excitement; increases heart rate; increases delivery of blood/glucose/oxygen to brain/muscles; prepares body for 'flight or fight' response.

   Allow other valid responses, e.g. diverts blood to muscles/increases sweating/(in animals) causes hairs to stand on end. [4 marks]

4. **a** Rate of energy release/respiration (in the body) at rest. [1 mark]

   b Any **one** from: slow growth; tiredness/fatigue. [1 mark]

Plant hormones

1. **a** Roots have (all) grown downwards; towards gravity. [2 marks]

   b (Positive) Gravitropism/geotropism. [1 mark]

c Any **one** from: anchorage; obtain water; obtain mineral ions. [1 mark]

d So growth/results not affected by light/phototropism. [1 mark]

2. a (Positive) Phototropism. [1 mark]

b Obtain light/for photosynthesis. [1 mark]

c Auxin collects on shaded side; causes increased cell growth/elongation (**not** cell division). [2 marks]

3. a Gibberellin. [1 mark]

b Ethene. [1 mark]

Uses of plant hormones

1. As a weed killer – auxin; to promote flowering – gibberellin; to help plant cuttings grow roots – auxin. [3 marks]

2. a Both axes with suitable scales and correctly labelled, with Ethene concentration on horizontal axis and Percentage of bananas on vertical axis; points plotted correctly (within a half-square accuracy); valid line of best fit. [3 marks]

b As the ethene concentration increases up to 4%, the percentage of bananas that are ripe (after three days) increases; as the ethene concentration increases above 4%, the percentage of bananas that are ripe (after three days) stays constant/decreases slightly. [2 marks]

c Any **one** from: age of bananas; type of bananas; temperature. [1 mark]

d Make other fruit go off; because they release ethene. [2 marks]

> **Section 6:** Inheritance, variation and evolution

Sexual and asexual reproduction

1. a Asexual; mitosis; genes. [3 marks]

b Only one parent/two parents not involved; no fusion of gametes/no mixing of genetic information. [2 marks]

c Different environmental conditions/named example, e.g. they will get different amounts of light/water. [1 mark]

2. a Fertilisation/fusion/joining of gametes/sex cells. [1 mark]

b Mixing of genes/DNA/genetic information; from both parents/from donkey **and** horse. [2 marks]

3. Female: (egg cells); egg cells; Male: pollen; sperm. [3 marks]

4. Worked example – details given in workbook.

Cell division by meiosis

1. a 24 [1 mark]

b 48 [1 mark]

c 48 [1 mark]

2. 3, 2, 1 [1 mark]

3. Similarity: contain same/half number of chromosomes **or** contain one copy of each chromosome pair/contain one copy of each chromosome. Difference: contain different versions of each chromosome **or** contain different alleles **or** are not genetically identical. [2 marks]

4. Worked example – details given in workbook.

DNA, genes and the genome

1. 4, 2, 1, 3 [1 mark]

2. Any **one** from: genes are (short) sections of/made of DNA; not all DNA is part of a gene; genes code for proteins but not all DNA codes for proteins. [1 mark]

3. Genes code for amino acid sequences; amino acids make up proteins; proteins are used for growth/structure/enzymes. [3 marks]

4. The entire genetic material of an individual/organism. [1 mark]

5. To look for genes linked to different diseases; understand/treat inherited disorders; trace human migration patterns. [3 marks]

Structure of DNA

1. a A = sugar; B = phosphate. [2 marks]

b Nucleotide. [1 mark]

c (Molecule) made up of repeating sub-units. [1 mark]

d Made of two strands; in a spiral. [2 marks]

2. a TACATGGAT All A and T paired up; all C and G paired up. [2 marks]

b 3 [1 mark]

c Gene. [1 mark]

3. 32; (32); 18; 18 [3 marks]

Protein synthesis and mutations

1. a 6, 2, 5, 1, 4, 3. Three or four or five in correct sequence = [1 mark] **but** all correct = [2 marks]

b Shape is necessary for protein function; valid example, e.g. shape of enzyme active site/shape of antibodies allows binding to antigens/protein structure gives collagen fibres their strength. [2 marks]

Level 2 [3–4 marks] Clear explanation linking a change in base sequence to a change in haemoglobin function.
Level 1 [1–2 marks] Partial/incomplete explanation linking a change in base sequence to a change in haemoglobin function.
No relevant content [0 marks]

Indicative content:

- Haemoglobin is a protein.
- Protein/haemoglobin coded for by DNA/base sequence.
- DNA/base sequence codes for amino acid sequence in protein/haemoglobin.
- Single base change alters DNA code.
- This alters an amino acid in the protein/haemoglobin.
- This alters the final protein/haemoglobin structure/shape.
- The altered structure/shape affects protein/haemoglobin function/ability to carry oxygen.

3. a Does not code for a protein. [1 mark]

b Switches other genes on/off. [1 mark]

Inherited characteristics

1. a Different version of a gene. [1 mark]

b Dominant alleles are always expressed (if present); recessive alleles are only expressed if two copies present/if no dominant allele present. [2 marks]

c Both can taste PTC. [1 mark]

d Both must be **Tt**; because one of their children is a non-taster/**tt**. [2 marks]

e

		Mother's gametes	
		T	t
Father's gametes	T	TT	Tt
	t	Tt	tt

Parental alleles correct; offspring genotypes correct. [2 marks]

f Identification of child who cannot taste as **tt**; probability = 1 in 4/25%/¼/0.25. [2 marks]

g No (no mark) (because for this couple the actual proportion is 1 in 2/50%/½/0.5); genetic diagrams/Punnett squares only show probability. [1 mark]

2. a bb [1 mark]

b

Level 2 [3–4 marks] Clear explanation supported by appropriate Punnett square diagrams.
Level 1 [1–2 marks] Partial explanation supported by appropriate Punnett square diagrams.
No relevant content [0 marks]

Indicative content:

- Cross with a bb/brown mouse.
- If any offspring are brown then black parent must be heterozygous/Bb.
- If no brown offspring/all offspring are black, then black parent is (probably) homozygous/BB.

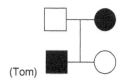

		Brown	
		b	b
Black	B	Bb	Bb
	b	bb	bb

half offspring brown half black

		Brown	
		b	b
Black	B	Bb	Bb
	B	Bb	Bb

all offspring black

Inherited disorders

1. a

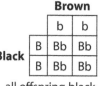

(Tom)

Correct except for 1 error = [1 mark] **but** all correct = [2 marks]

b Tom = Dd; mother = Dd; father = dd; sister = dd. [4 marks]

2. Worked example – details given in workbook.

Sex chromosomes

1. XY; 46; 23; meiosis. [4 marks]

2. a Nucleus. [1 mark]

b 150 million/half of them. [1 mark]

3. a

		Man	
		X	Y
Woman	X	XX	XY
	X	XX	XY

[1 mark]

b 0.5/½ /50%/1 in 2 [1 mark]

c 0.5/½ /50%/1 in 2 [1 mark]

Variation

1.

Characteristic	Genes only	Environment only	Both genes and environment
Blood group	✓		
Eye colour	✓		
Height			✓
Language spoken		✓	
Scar		✓	
Skin colour			✓

1 mark for each two correct ticks. Maximum = [3 marks]

2. a Has inherited genes/DNA from parents. [1 mark]

b Any **two** from: has inherited a mix of genes/DNA from both parents; environmental effects/named environmental effect, e.g. more food; mutation. [2 marks]

3. Worked example – details given in workbook.

Theory of evolution

1. (Over) three billion; simple(r); more advanced/modern. [3 marks]

2. a Any **four** from: (ancestors showed) variation in neck length; animals with longer necks were able to get more food; animals with longer necks therefore were more likely to survive and breed; passing on long(er) necks to offspring **or** passing on genes/alleles for long(er) necks; over time the average neck length increased **or** genes/alleles for long(er) necks became more common. [4 marks]

b Worked example – details given in workbook.

c (We now know that) environmental/acquired characteristics are not inherited; because the genes/DNA are not affected **or** neck length is controlled by genes/DNA. [2 marks]

3. The theory challenged the idea that God created all living things; there was insufficient evidence at the time the theory was published to convince all scientists; the mechanism of genes/inheritance/variation was not known in Darwin's time. [3 marks]

Darwin and Wallace

1. Natural selection. [1 mark]

2. Individuals with characteristics more suited to their environment are more likely to survive and reproduce. [1 mark]

3. a (On the) Origin of species. [1 mark]

b He knew there would be a lot of controversy; he wanted to collect more evidence. [2 marks]

4. Animals with warning colours are more likely to survive (predation); colours/genes/alleles for warning colours are passed on to offspring; so warning colours become more common in species. [3 marks]

Speciation

1. Organisms of the same species/Alsatians and Border Collies can interbreed producing fertile offspring; organisms of different species/Alsatians and wolves cannot produce fertile offspring. [2 marks]

2.

Level 3 [5–6 marks] Detailed explanation of how separation leads to independent evolution **and** speciation.
Level 2 [3–4 marks] Partial explanation of how separation leads to independent evolution **and** speciation.
Level 1 [1–2 marks] Some relevant points made about separation or independent evolution **or** speciation.
No relevant content [0 marks]

Indicative content:
- When water levels fall, separate lakes form and populations (of the same original species) are separated and can no longer interbreed.
- Different mutations occur in each population.
- Different environmental conditions in each lake.
- Natural selection selects for different characteristics in different lakes.
- Populations evolve independently in different lakes and develop different phenotypes.
- When water levels rise, lakes join together and populations (of same original species) can now mix.
- However, they have become so different they can no longer interbreed successfully/to produce fertile offspring.
- Separate species have formed/speciation has occurred.
- This process is repeated many times.

3. Unable to breed (extinct) animals together; so scientists cannot tell whether the animals could produce fertile offspring; individuals of the same species can show a lot of variation. [3 marks]

The understanding of genetics

1. Mendel carried out breeding experiments on plants; Mendel was the first person to discover genes, although he called them 'units'. [2 marks]

2. 2, 3, 1, 4. Three in correct order = [1 mark] **but** all correct = [2 marks]

3. a

	Pure-bred tall	
	T	T
t	Tt	Tt
t	Tt	Tt

Pure-bred short (row labels), Pure-bred tall (column labels)

	Tall	
	T	t
T	TT	Tt
t	Tt	tt

Tall (row labels), Tall (column labels)

Correct parental genotypes for first cross; correct offspring genotypes for first cross; correct parental genotypes for second cross; correct offspring genotypes for second cross. [4 marks]

b Plants have more offspring than mice; results are more repeatable/reliable. [2 marks]

4. Any **two** from: (at the time) no one knew about chromosomes/genes/DNA/ that chromosomes/alleles occur in pairs; scientists had other theories about inheritance, e.g. blending inheritance; Mendel's (mathematical) approach was novel; Mendel was not part of the academic establishment; his work was published in obscure journals/lost for many years; peas gave unusual results in comparison to other species. [2 marks]

Evidence for evolution

1. Development of antibiotic resistance in bacteria; Fossils; Similarities between living species. [3 marks]

2. a Chimpanzees and bonobos; have the most recent common ancestor. [2 marks]

b Worked example – details given in workbook.

3. a Remains of organisms from millions of years ago; **plus** any **three** from: hard parts/bones/shells **or** parts that do not decay/rot; parts replaced with minerals; bodies/parts that are preserved in resin/amber/ice/peat; footprints/burrows/root(let) traces. [4 marks]

b Any **two** from: many species did not live/die in environments suitable for forming fossils; many fossils not yet found; many fossils formed have been destroyed, e.g. by erosion/the rock cycle. [2 marks]

4. a

Level 3 [5–6 marks] Detailed explanation covering mutation **and** antibiotic resistance **and** natural selection.
Level 2 [3–4 marks] Partial explanation covering mutation **or** antibiotic resistance **and** natural selection.
Level 1 [1–2 marks] Partial explanation covering mutation **or** antibiotic resistance **or** natural selection.
No relevant content [0 marks]
Indicative content: • Bacteria develop (random) mutations.   • Introduces new strains into population.   • Some strains less affected by particular antibiotics/are antibiotic resistant.   • Antibiotics kill non-resistant strains/resistant strains are selected for.   • Reduced competition.   • Resistant strains survive and reproduce.   • New strain becomes more common in the population.

b Any **two** from: do not overuse antibiotics/do not use when inappropriate, e.g. for non-serious/viral infections; patients to complete courses of antibiotics; restrict agricultural use of antibiotics. [2 marks]

Extinction

1. a Extinct means that there are no remaining/living individuals left; endangered means that the numbers left are very low. [2 marks]

b Any **one** from: poaching; disease; lack of suitable habitat. [1 mark]

c Any **three** from: move them to a reserve/have guards to stop poachers; put in zoos/captive breeding programmes; educate people to stop poaching; use reproductive technology, e.g. place embryos in surrogate mothers of other types of rhinoceros. [3 marks]

Selective breeding

1. Any valid characteristic and suitable explanation, e.g. a gentle nature; so will not cause harm **or** obedient; so can follow instructions. [2 marks]

2. Any **three** from: large size; sweet taste; bright colour; juiciness/crispness; short/early ripening time; large number produced per tree; resistance to pests/disease; other valid answer. [3 marks]

3. a Choose parents who are fast runners; breed them together; from the offspring select the fastest runners; use these for breeding; repeat over many generations. [5 marks]

b Inbreeding; offspring prone to disease/inherited defects; loss of variation. [3 marks]

Genetic engineering

1. Transferring a gene from one organism into a different organism. [1 mark]

2. With genetic engineering you can transfer a gene/characteristic from any other species; with selective breeding you can only bring together characteristics/alleles within the same species. [2 marks]

3. Any **three** from: resistance to disease; resistance to pests; resistance to drought; resistance to cold/frost; resistance to herbicides/weed killers; bigger fruit; any other valid answer. [3 marks]

4. a Worked example – details given in workbook.

b

Level 3 [5–6 marks] Clear, logical and detailed description including correct references to enzymes and vector.
Level 2 [3–4 marks] A partial description omitting some details, but still giving a logical sequence.
Level 1 [1–2 marks] Some details correctly described.
No relevant content [0 marks]
Indicative content: • Insulin gene cut from human DNA/chromosome.   • Using enzymes/restriction enzymes.   • Use of a vector/plasmid.

- Vector/plasmid cut open using same enzymes/ restriction enzymes.
- Insulin gene inserted into vector/plasmid.
- Using another enzyme/ligase enzyme.
- Vector/plasmid inserted into bacterial cell.
- Bacteria reproduce producing many cells all containing human insulin gene.

Note – correct names of enzymes (restriction/ ligase) are **not** needed for full marks.

Cloning

1. a 1, 4, 5, 2, 3. Three or four in correct sequence = [1 mark] **but** all correct = [2 marks]

b Can produce more offspring; because use (several/ many) surrogates. [2 marks]

2. a Sexual reproduction; offspring will show variation/ may have different coloured flowers from parent. [2 marks]

b

Level 2 [3–4 marks] Detailed answer giving at least **one** advantage **and one** disadvantage for each method.
Level 1 [1–2 marks] Partial answer giving either advantages for both methods **or** disadvantages for each method **or one** advantage and **one** disadvantage.
No relevant content [0 marks]

Indicative content:

Advantages of tissue culture:
- offspring are genetically identical/clones of parent plant
- will have same colour flowers (as parent plant)
- can produce large numbers of plants.

Disadvantages of tissue culture:
- complicated/time-consuming process
- requires special equipment/materials, e.g. growth media/sterile conditions.

Advantages of taking cuttings:
- offspring are genetically identical/clones of parent plant
- will have same colour flowers (as parent plant)
- new plants will grow (relatively) quickly
- easy process.

Disadvantage of taking cuttings:
- can only get a limited number of cuttings from the parent plant.

3.

Level 3 [5–6 marks] Clear, logical and detailed description.
Level 2 [3–4 marks] A partial description with some details omitted or incorrect, but still giving a logical sequence.

Level 1 [1–2 marks] Some details correctly described.
No relevant content [0 marks]

Indicative content:
- Take skin cell from zorse.
- Take (unfertilised) egg cell from horse/zebra.
- Remove nucleus from egg cell.
- Nucleus from skin cell inserted into egg cell.
- Electric shock stimulates egg cell to divide to form an embryo.
- Embryo inserted into womb/uterus of female (horse/zebra/zorse).

Classification of living organisms

1. a (Carl) Linnaeus. [1 mark]

b Their structure/characteristics. [1 mark]

c 3, 5, 6, (1,) 4, 2, 7. 3/4/5 in correct sequence = [1 mark] **but** all correct = [2 marks]

d Binomial (system). [1 mark]

e *Homo habilis* [no mark] because it is in the same genus as humans. [1 mark]

2. a Any **two** from: animals; plants; fungi. [2 marks]

b Archaea; bacteria. [2 marks]

c Chemical analysis/RNA sequencing. [1 mark]

Section 7: Ecology

Habitats and ecosystems

1. a A population is all the individuals of the same species, e.g. all the frogs; the community is made up of all the populations, i.e. all the living organisms in and around the lake; a habitat is where a particular species lives, e.g. the habitat of the fish is the pond (water); an ecosystem is made up of all the living organisms together with the places they live, i.e. the whole community and the different habitats they live in. [4 marks]

b

Level 3 [5–6 marks] Explanation of possible changes in the ecosystem together with an appreciation that interdependence is complicated and it is impossible to predict changes with certainty.
Level 2 [3–4 marks] Explanation of some possible changes in the ecosystem.
Level 1 [1–2 marks] Description of some possible changes in the ecosystem.
No relevant content [0 marks]

2. A community in which population sizes stay fairly constant. [1 mark]

Food in an ecosystem

1. **a** Zebra/impala/giraffe. [1 mark]

   b Grass/acacia tree. [1 mark]

   c Cheetah. [1 mark]

   d Lion; predator/carnivore with no predators. [2 marks]

   e Cheetah/lion. [1 mark]

   f grass → zebra → cheetah → lion **or** grass → impala → cheetah → lion **or** acacia tree → impala → cheetah → lion

   Correct sequence; correct arrows. [2 marks]

   g From the Sun/light; by photosynthesis; by producers/first trophic level/(green) plants; which make glucose/biomass. [4 marks]

Abiotic and biotic factors

1.

Factor	Biotic	Abiotic
Disease	✓	
Food availability	✓	
Moisture level		✓
Predators	✓	
Soil pH		✓
Temperature		✓

Two or three correct = [1 mark] **but** four or five correct = [2 marks] **but** all correct = [3 marks]

2. Food; mates; territory. [3 marks]

3. To avoid competition with trees; for light. [2 marks]

4. Worked example – details given in workbook.

Adapting for survival

1.

Adaptation	Structural	Behavioural	Functional
Acacia trees...	✓		
Camels...			✓
Eagles...	✓		
Swallows...		✓	
Weaver birds...		✓	

1 mark for each tick. Maximum = [5 marks]

2. Live in extreme environments; example of extreme environment, e.g. high temperature/high pressure/high salt concentration. [2 marks]

3. Stomata needed for exchange of gases/oxygen/carbon dioxide; lower surface in water lily under water **or** stomata usually on lower surface as a way of reducing/controlling water loss/transpiration; do not need to conserve water as it is always available. [2 marks]

4. Being poisonous deters predators; bright colours warn predators of poison. [2 marks]

Measuring population size and species distribution

1. **a** Transect (line). [1 mark]

   b Worked example – details given in workbook.

   c Any **three** from: the percentage cover increases the further away from the tree; reaches a maximum 4 m from the tree; grass has to compete with tree for light/water/mineral ions; less competition further away from tree/reverse argument. [3 marks]

2. **a** Mean number of dandelions per quadrat = 8/10 = 0.8; mean number of dandelions per m² = 4 × 0.8 = 3.2; estimated number of dandelions in park = 3.2 × 120 = 384. [3 marks] Note – correct final answer with no working shown = [3 marks]

   b To avoid bias. [1 mark]

   c Random coordinates; produced by random number tables/random number generator. [2 marks]

   d Taken more readings (before calculating mean)/used larger quadrats. [1 mark]

Cycling materials

1. **a** W = photosynthesis; X = respiration; Y = respiration; Z = feeding/eating. [4 marks]

   b Combustion; of wood/fossil fuels; decay/decomposition; by microorganisms/bacteria/fungi/decomposers. [4 marks]

2. Evaporation (from seas/lakes/etc.); transpiration (from plants); condensation (as water vapour); precipitation (as rain). [4 marks]

3. Only a finite amount of materials on Earth; recycling provides materials necessary for life and growth. [2 marks]

Decomposition

1. a Lipids (in milk) decay/breakdown; forming fatty <u>acids</u> (and glycerol). [2 marks]

   b 45 (°C) [1 mark]

   c $1/27 = 0.037$; $= 3.7 \times 10^{-2}\,(s^{-1})$ [2 marks] Note – correct final answer with no working shown = [2 marks]

   d Up to 45 °C as the temperature increases so the rate of reaction increases; as temperature increases particles have more energy/rate of collision increases; above 45 °C the rate of reaction decreases; enzyme/lipase is becoming denatured/shape of active site is changing. [4 marks]

2. a Kitchen waste/agricultural waste/sewage. [1 mark]

   b Anaerobic/no oxygen; warm; moist/water present. [3 marks]

   c Aerated/oxygen present/not anaerobic. [1 mark]

Changing the environment

1. Pigeons… Human interaction; Pink-footed geese… Seasonal; Surtsey… Geographic. One or two correct = [1 mark] **but** all correct = [2 marks]

2. a They have less time to hunt on the ice; they catch less food. [2 marks]

   b They may move further north; where the sea ice is permanent/lasts longer **or** they may move further south/inland; to find other food. [2 marks]

3. Worked example – details given in workbook.

Effects of human activities

1. a The variety of all the different species within an ecosystem/on Earth. [1 mark]

   b To ensure the stability of ecosystems; by reducing the dependence of one species on another. [2 marks]

2. Human population is increasing; increased standard of living; more resources used; more waste produced. [4 marks]

3. Any **two** from: (provide more land) for cattle; (provide more land) for crops/biofuels/rice fields; to obtain wood, e.g. for building/furniture. [2 marks]

4. a (Garden) compost/fuel. [1 mark]

   b

Level 2 [3–4 marks] Description and explanation of some effects of using peat.
Level 1 [1–2 marks] Description of some effects of using peat.
No relevant content [0 marks]

Indicative content:

- Burning peat releases carbon dioxide.
- Peat, when it decays, releases carbon dioxide.
- Due to respiration by decomposers/microorganisms.
- Increased carbon dioxide emissions lead to global warming.
- Peat bog habitats support a variety of plant/animal/microorganism species.
- Removal of peat reduces (area of) peat bog habitats.
- Reduces biodiversity.

Global warming

1. Any **two** from: carbon dioxide; methane; water vapour; other valid example. [2 marks]

2. a To have a small(er) vertical scale; so small changes become more obvious. [2 marks]

   b

Level 3 [5–6 marks] Description of what the graph shows **plus** an evaluation of whether it provides evidence for global warming, with points made both in support and against.
Level 2 [3–4 marks] Description of what the graph shows **plus** some consideration of whether it provides evidence for global warming.
Level 1 [1–2 marks] Description of what the graph shows.
No relevant content [0 marks]

Indicative content:

- (Overall) the graph shows an increase in (global) temperature…
- of (about) 0.7 °C.
- This change may or may not be significant.
- There are many fluctuations/times when temperature falls.
- Graph shows temperature (differences) over more than a century…
- which should show a clear trend.
- However, it does not show how temperatures have changed further in the past…
- which could show a different overall trend…
- and does not show temperature changes since 2006.

3. Any **two** from: habitat loss; species migration/change in distribution; decrease in biodiversity/species extinction; any other specific valid answers [2 marks]

Maintaining biodiversity

1. a Increases numbers of the endangered species; for reintroduction into the wild. [2 marks]

   b Less land needed for landfill; wild habitats can be protected. [2 marks]

c Woodland contains a variety of plant life; provides food/shelter for many animal species. [2 marks]

d Rare habitats contain their own unique species; protecting the habitats also protects these species. [2 marks]

2. a Worked example – details given in workbook.

b Any **two** from: they provide a habitat for a wide range of plant and animal species; provide wildlife corridors, allowing wildlife to move freely between habitats to find food/shelter/mates; wildflowers are important sources of nectar and pollen for insect-pollinators so field margins/hedgerows promote the pollination of plant species dependent on insect-pollinators. [2 marks]

Biomass in an ecosystem

1. (Dry) mass of biological material. [1 mark]

2. a Drawing of a pyramid in the correct order with grass at bottom and labelled correctly; drawing should show the levels to scale. [2 marks]

b $160/2200 \times 100 = 7.3(\%)$; $20/160 \times 100 = 12.5(\%)$ [2 marks]

c Any **two** from: not all of the organisms in one trophic level are eaten by the next trophic level; not all ingested material is absorbed/some is egested as faeces; some material respired/lost as carbon dioxide/water; some materials lost as waste/urea/urine. [2 marks]

d Transfer from grass to rabbits is less efficient because grass is more difficult to digest/more is lost as faeces **or** transfer from rabbits to foxes is more efficient because meat is easier to digest/less lost as

faeces; plants are more difficult to digest because of plant cell walls/cellulose. [2 marks]

3. Not enough energy/biomass in the last trophic level to support another level. [1 mark]

Food security

1. Having enough food to feed a population. [1 mark]

2. Being at war; drought. [2 marks]

3. (Fishing quotas:) not all fish are caught/some fish remain; so remaining fish can breed (to restore fish numbers); (Control of net size:) young/small fish are not caught; so, they are able to breed (to restore fish numbers). [4 marks]

4. a Worked example – details given in workbook.

b Any **two** from: cruel to animals; product is not as healthy/may contain antibiotics; product is not as tasty. [2 marks]

Role of biotechnology

1. Does not contain meat so can be eaten by vegetarians/vegans/religious groups who cannot eat particular meats; lower in fat than meat so is seen as a healthier alternative. [2 marks]

2. a Contains gene(s) from another organism. [1 mark]

b Rice is a common/staple food; in the countries where vitamin A deficiency is most common. [2 marks]

3. a Use of living things/biological processes; to manufacture products. [2 marks]

b Human population is growing; demand for food will increase. [2 marks]